Biosecurity

Biosecurity

Understanding, Assessing, and Preventing the Threat

Edited by

RYAN BURNETTE

WILEY

Cover design: John Wiley & Sons, Inc.
Cover image: © Joseph Akers/courtesy of the author

Library of Congress Cataloging-in-Publication Data:

Biosecurity : understanding, assessing, and preventing the threat / edited by Ryan Burdette.
pages cm
Includes index.
ISBN 978-0-470-61417-4 (hardback)
1. Biosecurity. 2. Bioterrorism—Prevention. 3. Biological weapons—Safety measures. I. Burnette, Ryan, 1977– author, editor of compilation.
JZ5865.B56B58 2013
363.325'3—dc23 2013016290

Printed in the United States of America.

10 9 8 7 6 5 4 3 2 1

Contents

Preface xi

Author Affiliation. xiii

Acknowledgments xvii

About the Author. xxi

PART I **An Introduction to Biosecurity** 1

CHAPTER 1 Defining Biosecurity and Related Concepts 3

Ryan N. Burnette, Jenna E. Hess, Joseph P. Kozlovac, and Jonathan Y. Richmond

What is Biosecurity? 3

Related Concepts 4

Risk Management and Prioritization in Biosecurity 12

Basic Components of Biosecurity .13

Conclusion 14

CHAPTER 2 Biosecurity as a Function of Biosafety Microbiological Laboratories 17

Jonathan Y. Richmond, Ryan N. Burnette, and Gigi Kwik Gronvall

Introduction 17

Evolving Steps towards Biosecurity 18

Biosecurity is Embedded in Biosafety Programs 20

Control Access to High-Consequence Pathogens and Toxins 22

Accountability 22

Occupational Health Programs 23

Biosafety Culture 23
Conclusion 24

CHAPTER 3 A Perspective of Biosecurity: Past to Present 27
Robert J. Hawley and Joseph P. Kozlovac
Introduction 27
Biological Warfare 28
Bioterrorism 30
Biocrime 34
Bioaccidents 37
Conclusion 41

PART II **ELEMENTS OF BIOSECURITY** 49

CHAPTER 4 Physical Elements of Biosecurity 51
Jeff Slayton, Daniel N. Apple, and Benjamin A. Perman
Introduction 51
Protecting and Securing Biological Materials 51
Breaking Down the Program: The Six Security Capabilities 58
Architecture, Engineering and Work Practice Considerations 63
Operations and Sustainment 67
Conclusion 70

CHAPTER 5 Operational Elements of Biosecurity 71
J. Craig Reed and Debra C. Sharpe
Introduction 71
The Biosecurity Process 72
Biosecurity Challenges Associated with Different Institutions 82
New Regulatory Requirements 84
Conclusion 87

CHAPTER 6 Basic Principles of Threat Assessment 89
Benjamin A. Perman, Lindsay T. Odell, Jason Griffeth, and Daniel N. Apple
Introduction 89
A Comprehensive Approach to Biosecurity 89
Safety versus Security: Risk vesus Threat 90

The Purposes and Requirements of Basic Suitability and Threat Assessment Programs 102
Conclusion 108

PART III **BIOSECURITY IN VARIOUS SECTORS** 111

CHAPTER 7 Biosecurity in Research Laboratories 113
Kavita M. Berger
Introduction 113
Laboratory Biosecurity 114
The Select Agent Program 115
Conclusion 122

CHAPTER 8 Biosecurity in the Food and Agricultural Industries 129
Henry S. Parker
Introduction 129
The U.S. Food and Agriculture Infrastructure 129
Food Defense 132
Food Defense: Protecting U.S. Food and Agriculture from Intentional Disruption 145
Conclusion 150

CHAPTER 9 Biosecurity in Mass Transportation Networks 155
Brian K. George
Introduction 155
Mass Transit Systems and Vulnerabilities 155
Federal, State, and Local Biosecurity Activities and Planning .157
Conclusion 164

CHAPTER 10 Utilizing Biosecurity Principles to Combat Naturally Occurring Epidemics 167
J. Craig Reed, Robert A. Heckert, Patricia Delarosa, and Vera Ettenger
Introduction 167
Separating Deliberate Outbreaks from Naturally Occurring Epidemics 168
The Need for Biosafety and Biosecurity in the Response to a Naturally Occurring Outbreak 169

Assessing Biosafety and Biosecurity Risks in Public Health: The Science of Biosecurity 172

Increasing Security Functions through the Application of Biosecurity to Public Health 174

Challenges Associated with the International Coordination of Outbreak Response. 175

Biosecurity Impediments to International Health Programs 178

Conclusion 180

PART IV **BIOSECURITY POLICY, BIOTERRORISM, AND THE FUTURE** 185

CHAPTER 11 Current Efforts to Fortify Biosecurity Policy in the U.S. 187

Gigi Kwik Gronvall

Introduction 187

Emerging Policies Limiting Access to Pathogens and Enhancing Laboratory Security. 188

Policies that Focus on Dual-Use Research of Concern 191

Evolving Oversight of Synthetic Biology 192

Conclusion 194

CHAPTER 12 The Link between Biosecurity and Bioterrorism. 197

Daniel N. Apple and Benjamin A. Perman

Introduction 197

Bioterrorism Attack Cycle. 197

The Bioterrorism-Biosecurity Nexus. 202

The Biological Criminal Continuum 204

Conclusion 206

CHAPTER 13 The Importance of Engagement and Education for Effective Biosecurity 209

James Revill and Catherine Jefferson

Introduction 209

Overview of Engagement and Education 209

Other Intervention Points and Actors. 213

Challenges Moving Forward 215

Why Bother with Biosecurity-Related Education and Engagement? 216
Moving Biosecurity Engagement and Education Forward 218
Conclusion 221

CHAPTER 14 Monitoring and Evaluating Laboratory Biosafety Programs. 225
Heather N. Meeks, Betiel H. Haile, Ngozi Erondu, Lisa Ferland, Meeyoung Park, Affan Shaikh, and Scott J.N. McNabb
Introduction 225
Need for a Laboratory Biosafety Program Framework. 227
Major Concepts and Definitions 230
Conceptual Framework. 239
Laboratory Biosafety Indicators. 241
Anticipated Benefits of the Indicator Database. 242
Analytic Processes to Refine Indicators 242
Limitations. 254
Conclusion 254

CHAPTER 15 The Future of Biosecurity: A Global Context 259
Ryan N. Burnette, J. Craig Reed, and Patricia Delarosa
Introduction 259
Biosecurity Summarized 259
A One World/One Health Model for Biosecurity 262
Future of Biosecurity. 266
Conclusion 267

APPENDIX 1 U.S. Food Defense since 9/11: Public Sector Initiatives and Programs 271
Henry S. Parker

APPENDIX 2 Timeline of Selected Events in Biosecurity Policy 289
Gigi Kwik Gronvall

Index. 299

Preface

The term "biosecurity" remained largely obscure to the public until the events of September 11, 2001, followed by the anthrax letters in the same year. Biosecurity, and the concepts it embodies, however, date back well before those fateful events. Yet those events initiated a wave of awareness and concern in the minds of the public and our legislators. Changes took place more rapidly than they ever had before: a new cabinet-level office, the Department of Homeland Security, was formed. Funding of biodefense programs grew exponentially. Stockpiles of vaccines were created. The hunt for novel vaccines and therapeutics against exotic pathogens increased rapidly. In short, the U.S. and much of the world began to prepare for similar events that were believed to be imminent. Despite this surge in activity and awareness, biosecurity remains poorly understood.

As this book will explore, biosecurity is a complex matrix of terminology, concepts, ideals, practices, and policy. Further complicating the concept of biosecurity is the vast number of related terms: bioterrorism, biodefense, bioweapons, and the like are arguably more common in the modern vernacular than the word "biosecurity." In response, this book begins by disassembling the tenets of biosecurity, breaking it down into its fundamental pieces. Physical and operational biosecurity will be described and the concepts of assessing threats fully detailed. We will explore how biosecurity impacts and is impacted by a variety of industries distinct from the traditional research laboratory. Many of these industries, such as agriculture, mass transit, and public and global health, will receive new attention under the lens of biosecurity. This book will also take a detailed look at current biosecurity policy and where new policy is likely to develop. Potentially for the first time, this book discusses the importance of education and advocacy for biosecurity in frank terms, which may assist researchers in the integration of biosecurity into the educational process. This book also presents a novel tool developed to quantify the effectiveness of laboratory biosafety and biosecurity programs.

The intent of this book is to provide clarity to the term "biosecurity" and demonstrate its relevance to areas that have historically been given little consideration. Finally, this book challenges the reader to look forward, armed with new information about biosecurity, to how biosecurity as an ideal needs to develop and progress. This is not an easy task, but, as this book demonstrates, it is a necessary task.

It is my sincere hope that this book will do two things: provide the reader with a foundation for everything that biosecurity is and is not and compel the reader to think about the importance of biosecurity as it relates to our collective security, safety, and health.

Ryan N. Burnette, Ph.D.
December 21, 2012

Author Affiliation

CHAPTER 1 **Defining Biosecurity and Related Concepts**

Ryan N. Burnette, Ph.D.,[1] Jenna E. Hess, M.S.,[1]
Joseph P. Kozlovac, M.S., SM (NRCM), CBSP,[2]
Jonathan Y. Richmond, Ph.D., RBP[3]

[1] Alliance Biosciences

[2] Agency Biosafety Officer, USDA Agricultural Research Service (ARS), National Program Staff, Animal Production and Protection

[3] Jonathan Richmond & Associates

CHAPTER 2 **Biosecurity as a Function of Biosafety for Microbiological Laboratories**

Jonathan Y. Richmond, Ph.D., RBP,[1] Ryan N. Burnette, Ph.D.,[2]
Gigi Kwik Gronvall, Ph.D.[3]

[1] Jonathan Richmond & Associates

[2] Alliance Biosciences

[3] Center for Biosecurity, University of Pittsburgh Medical Center

CHAPTER 3 **A Perspective of Biosecurity: Past to Present**

Robert J. Hawley, Ph.D., RBP, SM (NRCM), CBSP,[1]
Joseph P. Kozlovac, M.S., SM (NRCM), CBSP[2]

[1] Alliance Biosciences

[2] Agency Biosafety Officer, USDA Agricultural Research Service (ARS), National Program Staff, Animal Production and Protection

CHAPTER 4 Physical Elements of Biosecurity

Jeff Slayton,[1] Daniel N. Apple, and
Benjamin A. Perman, Ph.D., PPS,[2]

[1] Gregg Protection Services; American Biological Safety Association, Texas Association of Hostage Negotiators

[2] Booz Allen Hamilton, Inc.

CHAPTER 5 Operational Elements of Biosecurity

J. Craig Reed, Ph.D., RBP,[1] and
Debra C. Sharpe, MPH, CCHO, RBP[2]

[1] Alliance Biosciences

[2] The Working Buildings Companies, LLC

CHAPTER 6 Basic Principles of Threat Assessment

Benjamin A. Perman, Ph.D., PPS,[1] Lindsay T. Odell, Ph.D.,[1]
Jason Griffeth,[1] and Daniel N. Apple

[1] Booz Allen Hamilton, Inc.

CHAPTER 7 Biosecurity in Research Laboratories

Kavita M. Berger, Ph.D.[1]

[1] Center for Science, Technology, and Security Policy, American Association for the Advancement of Science

CHAPTER 8 Biosecurity in the Food and Agricultural Industries

Henry S. Parker, Ph.D.[1]

[1] Georgetown University Medical Center, Department of Microbiology and Immunology, Washington, DC

CHAPTER 9 Biosecurity in Mass Transportation Networks

Brian K. George[1]

[1] The Tauri Group, LLC

CHAPTER 10 Biosecurity in Mass Transportation Networks

J. Craig Reed, Ph.D., RBP,[1] Robert A. Heckert, BSc(Agr), DVM, Ph.D., CBSP, SM(NRCM),[2] Patricia Delarosa, Ph.D., RBP, CBSP,[3] and Vera Ettenger, Ph.D., RBP, CBSP[3]

[1] Alliance Biosciences

[2] Robert Heckert Consulting

[3] Booz Allen Hamilton, Inc.

CHAPTER 11 Current Efforts to Fortify Biosecurity Policy in the U.S.

Gigi Kwik Gronvall, Ph.D.[1]

[1] Center for Biosecurity, University of Pittsburgh Medical Center

CHAPTER 12 The Link between Biosecurity and Bioterrorism

Daniel N. Apple and Benjamin A. Perman, Ph.D., PPS[1]

[1] Booz Allen Hamilton, Inc.

CHAPTER 13 The Importance of Engagement and Education for Effective Biosecurity

James Revill, Ph.D.,[1] and Catherine Jefferson, Ph.D.[2]

[1] Harvard Sussex Program, University of Sussex

[2] Kings College, London

CHAPTER 14 Monitoring and Evaluating Laboratory Biosafety Programs

Heather N. Meeks, Ph.D.[1], Betiel H. Haile, M.P.H.[1], Ngozi Erondu, M.P.H.[2], Lisa Ferland, M.P.H.[2], Meeyoung Park, M.P.H.[2], Affan Shaikh, M.P.H.[2], and Scott J.N. McNabb, Ph.D., M.S.[2,3]

[1] Defense Threat Reduction Agency, Basic and Applied Sciences

[2] Public Health Practice, LLC

[3] Emory University, Rollins School of Public Health

CHAPTER 15 The Future of Biosecurity: A Global Context

Ryan N. Burnette, Ph.D.[1], J. Craig Reed, Ph.D., RBP,[1] and Patricia Delarosa, Ph.D., RBP, CBSP[2]

[1] Alliance Biosciences

[2] Booz Allen Hamilton, Inc.

APPENDIX 1 U.S. Food Defense since 9/11: Public Sector Initiatives and Programs

Henry S. Parker, Ph.D.[1]

[1] Georgetown University Medical Center, Department of Microbiology and Immunology, Washington, DC

APPENDIX 2 Timeline of Selected Events in Biosecurity Policy

Gigi Kwik Gronvall, Ph.D.[1]

[1] Center for Biosecurity, University of Pittsburgh Medical Center

Acknowledgments

It is imperative to stress that nothing about managing and editing a volume such as this one is an independent exercise, and there have been many helpful hands to assist in guiding the process. At the same time as it goes without saying, the overwhelming majority of thanks must be given to the authors of and contributors to this manuscript. They have committed countless hours of their lives to crafting useful information that we all hope will be well received and well utilized. For their commitment to this project, I am eternally grateful. I can safely admit that I am nowhere near expert enough to have drafted the content that was generously contributed.

Jenna E. Hess, M.S., presented as an author and co-editor of this volume, was instrumental in the overall process. Her attention to detail and writing skills have resulted in a much better product than it would have been if left to my own devices. Her instincts are solid and were brought to bear in this project.

J. Craig Reed, Ph.D., RBP, is easily one of the greatest "team players" not only on my company's staff, but on this project as well. He contributed more content than any other single author, rolled with all the punches, and even provided editorial assistance on several chapters. Craig's viewpoints helped shape the way many of the topics were presented, even changing my own mind more than once.

Time is always a factor when drafting such a volume, and yet is it something we are all in need of. I was provided the time to manage this effort in a large part thanks to the understanding and patience of the owners of my company. Jim Burnette and Kevin Kokal have been tremendously supportive, both of my time requirements, and of the vision to see the effort to completion. I am grateful to both for many reasons beyond the scope of this book as well. In fact, this effort would not have been possible without their unwavering commitment to me as the company has grown. Had they not had confidence in me years ago, the opportunity to create this book would never have arisen.

My introduction to biosecurity truly began in the field of biosafety, which is discussed at length in this book. I owe a great deal of thanks (and admiration) to the mentors I have had in this field. A single phone call changed my career when Jonathan Y. Richmond, Ph.D., RBP, politely answered. And thus began my career in the world of biosafety and biosecurity. His steady, consistent mentorship has literally made me the professional I am today. I am flattered to be able to call one of the true fathers of the industry a friend. Likewise, Robert (Bob) Hawley, Ph.D., RBP, CBSP, SM(NRCM), and Robert Heckert, DVM, Ph.D., RBP, CBSP, SM(NRCM), have both spent considerable time providing me with education and insight. Marian Downing, RBP, CBSP, SM(NRCM), has been by my side, supporting and guiding me for more years than any one person in this field. She too has rolled with more punches than I ever thought I would throw, yet she is still dedicated and here. Christina Thompson, M.S., RBP, CBSP, SM(NRCM), referred me to Marian and therefore to a large part of who I am today. Chris remains a true team player, consultant, and friend.

Daniel N. Apple and Benjamin A. Perm would like to thank the following individuals for substantive contributions to the development of the materials and ideas presented in Chapter 6: Chuck Tobin; Dr. Ronald Schouten; Dr. Tricia De La Rosa, RBP, CBSP; Dr. Vera Ettenger, RBP, CBSP; Susan Weekly; Dr. Casey Skvorc; Supervisory Special Agent Edward You; and Dr. Nanda Gudderra. The authors also thank John Page for valuable editorial comments on the manuscript. We would also like to thank our friends in the Republics of Azerbaijan, Georgia, Kazakhstan, Ukraine, and Uzbekistan and the Russian Federation, who have taught us more about biosecurity than they may realize.

Virginia (Ginny) Howe has been there through the entire process, carefully answering my questions and guiding the entire process. I owe a great deal of thanks to her and the late Roger Woodson.

For more than one reason, I owe a great deal of thanks to Kady Dudley. She has been supportive of the process and the time involved and has remained by my side to see it through. Likewise, Jim and Melinda Burnette remain the origin of so many opportunities that ultimately led to this book.

Ryan N. Burnette, Ph.D.

December 27, 2012

About the Author

Ryan N. Burnette, Ph.D. is the Director of Alliance Biosciences, a laboratory design and biorisk management consulting firm headquartered in Virginia. Alliance provides laboratory design and biorisk management services to private industry, academia and government institutions in the U.S. and more than a dozen countries. In addition to serving as a biosafety and biosecurity consultant, Dr. Burnette has published in the fields of molecular biology, endocrinology and infectious disease. Dr. Burnette previously held positions in the Departments of Biology and Biochemistry at Virginia Tech, and the Department of Molecular Physiology and Biophysics at Vanderbilt University School of Medicine prior to creating Alliance Biosciences. For more information, please visit www.AllianceBiosciences.com.

PART I

AN INTRODUCTION TO BIOSECURITY

CHAPTER 1

Defining Biosecurity and Related Concepts

Ryan N. Burnette, Jenna E. Hess, Joseph P. Kozlovac, and Jonathan Y. Richmond

WHAT IS BIOSECURITY?

Biological security, or "biosecurity," is not easy to define and elicits a variety of interpretations. Thus, it is important to clearly define the context in which the term is used. In a broad sense, it is a "strategic and integrated approach, encompass[ing] the policy and regulatory frameworks that analyze and manage risks in the sectors of food safety, animal life and health, and plant life and health, including associated environmental risk."[1] However, biosecurity is not limited to policy and regulation, as this book demonstrates. For the purpose of defining biosecurity, this chapter focuses on elements of laboratory biosecurity.

Laboratory biosecurity is a [set of] concepts and practices used to secure sensitive biological materials from persons or entities that should not have access.[2] The World Health Organization (WHO) refers to "laboratory biosecurity ... [as the] institutional and personal security measures designed to prevent the loss, theft, misuse, diversion, or intentional release of pathogens and toxins."[3] Biosecurity is not strictly limited to biological agents or harmful byproducts. It also applies to products having intrinsic value, such as novel vaccines, biological therapeutics, information-technology platforms, synthetic nanoparticles or organisms, and products having high monetary value or related to biological agents. These will be referenced throughout this book as valuable biological material, or VBM.

Biosecurity as a field likely originated from its applications in controlling the release of genetically modified organisms, or GMOs, into indigenous environmental populations.[4] As with other condensed "bio" terms, the inclusion of the word "biosecurity" into popular vernacular was firmly established following the attacks of September 11, 2001. The Centers for Disease Control and Prevention (CDC) and the National Institute of Health (NIH) recognize that a robust biosafety program includes many facets of biosecurity and that it assumes many definitions.[5] For example, biosecurity in the animal industry is the protection of animal colonies from microbial contamination.

Biosecurity is a combination of terms. For example, the root is the all-familiar word, "security." The word "security" is also represented in the larger context of "safety and security." Therefore, safety, as a concept, can be thought of as a component of overall security. In fact, CDC often discusses biosecurity as a subset of a robust biosafety program.[5] The phrase "safety and security" is somewhat ubiquitous in our culture.

One of the most familiar examples of "security" is commercial air travel. In this sense, security is a set of measures to safeguard and protect travelers and commercial aircraft. Consider the extensive security procedures each passenger encounters prior to boarding the aircraft. While navigating these procedures may only take several minutes, their development and implementation took years of planning. It is important to consider the individual components of the airport security process.

You as a passenger encountered numerous personnel, each with different roles. Whether they work behind the scenes to screen checked luggage, verify photo identification and boarding passes, or operate the X-ray machinery, each plays a specific role. In addition, you were subjected to several highly specialized pieces of equipment, each with discrete functions: X-ray scanners, explosive residue sniffers, barcode scanners, manifest logs, metal detectors, and others. Taken alone, none of the components can be thought of as "airport security." But taken as a whole, and orchestrated through well-defined procedures and policy, the personnel and equipment constitute a comprehensive example of "security." This book assumes a similar approach to biosecurity. Like commercial air travel, biosecurity can be dissected into its finite parts and pieces.

RELATED CONCEPTS

Biosecurity has analogous elements to commercial air travel: physical, procedural, and personnel. It too is comprised of several, if not multiple, components that coalesce into the term "biosecurity." Further sections of this book will focus on the specifics of personnel, procedures, policy, and specialized equipment utilized. Major concepts related to biosecurity include:

- Biorisk
- Biohazard
- Biosafety
- Biocontainment
- Biosurety
- Biodefense
- Bioweapons
- Bioterrorism

Defining Biorisk

The WHO defines biorisk as "the probability or chance that a particular adverse event [e.g., accidental infection or unauthorized access, loss, theft, misuse, diversion, or intentional release], possibly leading to harm, will occur." From this data, a biorisk assessment is "the process to identify acceptable and unacceptable risks (embracing biosafety risks, risks of accidental infection) and laboratory biosecurity risks (risks of unauthorized access, loss, theft, misuse, diversion, or intentional release) and their potential consequences." Biorisk assessments are commonly used by laboratory programs to determine the level of risk a biological agent presents to the laboratory worker and strategies to mitigate those risks.

The management of biorisks (biorisk management) is defined as "the analysis and development of strategies to minimize the likelihood of the occurrence of biorisks."[6] Biorisk management requires the active participation of an institution's senior leadership (perhaps with the support of a biorisk committee) to recognize the institution's biorisks and to develop appropriate mitigation strategies as well as provide support and active leadership for the biorisk management program. The focus of attention should be on the high-consequence pathogens that have the greatest potential to negatively impact public health and agriculture and pose serious adverse economic consequences. Events can be naturally occurring, accidental, or deliberate. Therefore, provision of specific knowledge and practices prepares responsible partners to address the unexpected.[6]

Defining Biohazards

Literally, a biohazard is a hazard of biological origin. Such hazards could include infectious organisms, such as viruses and bacteria; noninfectious toxins, such as venoms and plant extracts; tissues and cultures, such as blood; vectors of disease, such as various arthropods; parasites, such as the malaria-causing Plasmodium parasite; and even therapeutics, such as insulin or Botox. Biohazards may present a risk to human, animal, plant, or environmental health. However, from this definition it is important to note that all infectious organisms and agents are considered biohazards, but not all biohazards are infectious organisms and agents.

Defining Biosafety

As referenced above, "safety" and "security" are two words often associated with each other. Thus, it is reasonable to infer that biosafety and biosecurity are components of a comprehensive biorisk management plan. Specifically, biosafety stands for biological safety. Chapter 2 will explore the concepts of biosafety in greater detail, but it can be defined as a set of principles and practices that dictate the safe handling and containment of potentially harmful biological agents for the purpose of preventing infection of laboratory workers and the public.[5]

In other words, biosafety encompasses an established set of guidelines allowing researchers, physicians, and other biomedical professionals to safely practice their work with harmful biological agents. It can be argued that biosecurity is actually a contributing philosophy to the overall concept of biosafety, since a breach of biosecurity could lead to a threat of infection to laboratory workers, the public, or agriculture. It is best to consider biosecurity and biosafety equally important conceptual contributors to the overarching concept of biorisk management. Despite significant overlap, biosafety and biosecurity remain distinct from one another: biosafety primarily deals with protection of the individual from the biological agent, whereas biosecurity deals with the protection of the biological agent itself.

Biosafety has been practiced to some degree since it was recognized that illnesses were passed from one entity to the next without the infectious substance being visible. The wearing of plague suits and masks by physicians during the Black Death in Europe is an example of a crude barrier to protect the healthcare worker. In 1974 the CDC published significant guidelines for working safely with various microbial agents.[7] Additional guidelines for improving biosafety were also published at that time.[8, 9, 10]

However, it was not until 1984 that a comprehensive approach was taken to draft a working document that encapsulated the critical components of safe handling and containment of potentially infectious microorganisms and toxins. The CDC/NIH publication Biosafety in Microbiological and Biomedical Laboratories (BMBL) has seen several revisions and is considered the code of safe practice for modern microbiological research.[5] Published by the Centers for Disease Control and Prevention and the National Institutes of Health (CDC/NIH), this document has led a global effort to codify the practices of biosafety. Equally important and influential is the World Health Organization Laboratory Biosafety Manual, one of the most recognized volumes of biosafety in the world.[3] These works spawned other efforts to codify these practices globally and are discussed in subsequent chapters. In short, the principles of biosafety have shaped how labs are designed, how experiments are conducted, and how biological agents are handled and manipulated.

Defining Biocontainment

Biocontainment, or biological containment, refers to the physical and procedural components of properly and safely keeping biological organisms and agents within the confines of some barrier system. This word is often used interchangeably with biosafety; although biocontainment usually implies some physical measure of isolation and containment. The term biocontainment is most commonly associated with microbiological laboratories and vivariums, where closed tubes, cages, biological safety cabinets, or the lab itself are considered to represent containment. But the term can be extrapolated to describe other types of containment as well.

Chapter 2 discusses in more detail the levels and criteria for biocontainment, but it is important to note there are usually two distinct "levels," primary and secondary. Primary biocontainment refers to the immediate protection of the laboratory worker from the laboratory environment. For example, standard manipulation of infectious microorganisms is often performed in biological safety cabinets (BSC), an engineering control designed for the safe manipulation of infectious materials. In a BSC, air is drawn into the cabinet, and microorganisms cannot move against the direction of air flow and escape from the BSC into the general lab environment. Therefore, a BSC is a form of primary biocontainment. Another example of primary biocontainment are the "fully encapsulated suits" associated with maximum containment facilities by which the lab worker is isolated from the microbial agents that may be present in the lab environment.

Secondary biocontainment usually refers to the lab itself and the building systems associated with the lab (HVAC, autoclaves, etc). Biocontainment labs have special engineering parameters designed to prevent the release of infectious agents, such as bacteria or viruses. While there are exceptions, most direct manipulation of infectious agents, such as the opening of a test tube or injection of a rodent, takes place in the primary barrier such as a BSC. In the event of a spill or other unintentional release from primary biocontainment, the lab also has design and operational components that act as a secondary form of biocontainment. Together, they can be thought of as two major levels of biocontainment aimed at the protection of the lab worker and the prevention of the escape of infectious agents into the general environment.

Defining Biosurety

Biosurety, or biological surety, is a word not as commonly heard as biosafety or biohazard. However, following the events of September 11, 2001 and the anthrax letter incident, biosurety has become more commonly used as a component of biosafety and biosecurity programs. Biosurety describes laboratory operations from the perspective of the operating environment and includes biosafety, physical security, personnel responsibility, and agent accountability. Biosurety often includes the accountability (inventory control) of the microorganisms, particularly select agents and toxins.[11]

The personnel reliability component of a biosurety program involves routine employee screening practices, such as conducting criminal and background checks on personnel who have direct access to infectious agents or the information pertaining to these agents. An overall goal of a biosurety program is to ensure that only individuals who have been adequately trained and can conduct work with high-consequence agents responsibly will have access to those agents.

In the U.S., the first biosurety programs were developed by the Army. They were adapted from other military surety programs, such as chemical and nuclear programs. The culmination of the Army's efforts resulted in the AR 50-1 policy on biosurety.[12] Other institutions, most notably those with maximum containment facilities, have developed their own version of comprehensive biosurety programs, such as those of the NIH and the University of Texas Medical Branch. These civilian agency programs have similar elements to the U.S. Department of Defense (DoD) programs, with a focus on fostering a culture of trust and personal responsibility. These programs typically have in place self-reporting mechanisms with which employees can and are encouraged to request to be temporarily removed from laboratory access based upon a variety of individual issues that prevent them from focusing adequately on the safe and secure conduct of high-risk laboratory work.[13,14]

Defining Biodefense

Biodefense, or biological defense, refers to an entity's policies and procedures to defend against an attack involving biological agents or weapons. Additionally, biodefense refers to a local community's emergency response efforts in the event of an attack or unintentional release. The terms biodefense and biosecurity are often and incorrectly used interchangeably. Strictly speaking, laboratory biosecurity is a set of procedures and policies that restricts access to potentially harmful biological agents. Biodefense, in simplistic terms, defines an entity's policies and procedures to defend itself beyond the laboratory environment from infectious and emerging disease outbreaks resulting from natural, accidental, or intentional introductions into susceptible populations. Biodefense strategies and actions can be used, however, to emphasize the need for the responsible conduct of science to include the use of good biosafety and laboratory security practices. While many governments worldwide have had certain biodefense policies in place since World War II, September 11, 2001 marked a turning point in the U.S. government's decision to fund biodefense efforts. Biodefense policies are scattered among many departments and agencies, such as the U.S. Department of Homeland Security (DHS) and DoD. Many other agencies have significantly contributed to the development of U.S. biodefense strategies and operations.

Defining Bioweapons

The loose definition of a "bioweapon" is a weapon that incorporates a biological agent. However, this definition must be narrowed. Several attributes are important to consider when defining a bioweapon.

Not all biological or infectious agents make suitable weapons. Therefore, one criterion used to evaluate the threat or risk of an infectious agent is its ability to be weaponized. For example, *Bacillus anthracis* is a spore-forming bacterium that causes the toxin-mediated disease state known as anthrax, which has three different pathologies (pulmonary, gastrointestinal, and cutaneous anthrax) based upon the route of exposure. Most notably, this agent made headlines with the "anthrax letters" of 2001, to be discussed in several subsequent chapters. Like all infectious biological agents, *Bacillus anthracis* is a naturally occurring bacterium found in the environment. Yet a handful of soil on a farm containing the bacterium is not "weapons-grade." Substantial manipulation of the bacterium, such as culturing, bulk production, and purification, is necessary to develop large quantities of *B. anthracis* spores. It is important to note, however, that biological agents do not necessarily have to be weaponized to create a successful biocrime or bioterrorist event. For example, the illicit application of naturally occurring *Salmonella* to a public salad bar requires no "weapons-grade" procedure but has the ability to adversely impact the public and can generate significant panic in the community. Only a relatively small number of people were actually killed as a result of the "anthrax letters."

However, the "attack" invoked widespread panic, significant mitigation, and investigation costs.

Although not universal, the following general criteria differentiate a naturally occurring infectious agent from a bioweapon. Can the infectious agent(s) be:

- Isolated
- Cultured
- Purified
- Scaled to production
- Adapted for delivery

The intent of a bioweapon is critical to define. Will the weapon be used to infect a small group of people, a large group of people, or maybe even a group that can infect others? Therefore, a "good" bioweapon is also defined by its transmissibility. Toxins, such as ricin or anthrax, only affect the exposed individual, and typically there is no secondary transmission. Consequently, even though ricin or anthrax may be suitable to infect a mass number of individuals, the effect is limited to those initially exposed. Despite the "small" number of potential deaths, this is still considered a bioterrorist event. As such, certain contagious viruses, meeting the criteria listed above, may make "good" bioweapons.[15]

Obviously, the goal of implementing a bioweapon is to cause harm or death to people, agriculture, or the environment or create economic disaster, the illusion of impending doom, or even simply to generate panic. Although not a terrorist event, a single smallpox infection in Yugoslavia (1972) quarantined nearly 10,000 people, closed country borders, and disrupted commerce. This illustrates how a small outbreak can have significant social and economic consequences.[16] Therefore, the sophistication of purifying and mass-producing weapons-grade anthrax-producing bacteria is not always necessary to define a bioweapon; simpler forms of bioweapons have been employed throughout history (Chapter 3). For example, in the late 1700s European soldiers gave smallpox-infected blankets to Native Americans in the French and Indian War. In this instance, smallpox was used as a bioweapon, and the inoculated blankets were the mechanism of delivery. At that time there was little knowledge of how bioweapons "worked"; however, this rudimentary attack caused devastating consequences to the indigenous populations.[15]

Conversely, entire laboratories and programs were intentionally developed to create new and more devastating bioweapons. In the former Soviet Union,

Vozrozhdeniya Island was established as a bioweapons research and manufacturing hub during the Cold War.[17] This sophisticated R&D and production facility existed solely to develop and manufacture potential bioweapons.

Defining Bioterrorism

The threat or use of biological agents as weapons to cause fear, terror, economic and political disruption, and unrest among the populace to achieve political, ideological, social, and/or religious goals is the hallmark of biological terrorism or bioterrorism. This definition is complicated by the potential for state-sponsored terrorism against strategic targets as well as the potential for using biological agents in the commission of crimes that do not have political goals but are committed for purposes of extortion or revenge.[18] Bioterrorism is a growing concern for governments around the world and has become a staple in best-selling fiction and Hollywood movies. At the root of bioterrorism is the word "terrorism." At its core, the goal of terrorism is not always the loss of large numbers of life. Rather, terrorism is successful if it has created terror or fear in large numbers of people. One definition of terrorism states: "criminal acts intended or calculated to provoke a state of terror in the general public, a group of persons, or particular persons for political purposes are in any circumstance unjustifiable, whatever the considerations of a political, philosophical, ideological, racial, ethnic, religious, or any other nature that may be invoked to justify them."[19]

Understanding the terrorist is a critical component to this discussion and explains why bioweapons are becoming more attractive to a variety of terrorist groups. There are generally five classes of terrorists: government-trained professionals, religious extremists, radical revolutionaries, mercenaries, and amateurs.[20] While each class of terrorist or terrorist organization may have disparate motives for eliciting terror, they all usually employ similar mechanisms of terror. We know there are a wide variety of terrorist mechanisms and weapons, from the commercial airliners used in the attacked on September 11, 2001, to suicide bombers in public places, to chemical, biological, radiological, and nuclear (CBRN) weapons.

Historically, bioweapons have not been used in many terrorist activities for a few reasons. The expertise required to develop bioweapons for terrorist agendas is extensive and in many cases unobtainable by less organized terrorist groups. Additionally, meeting the criteria to develop a "good" bioweapon is difficult. In short, there are few infectious agents that qualify as "good" bioweapons. Further, most agents that have been used in biocrimes or bioterrorist events have been food-borne pathogens requiring little manipulation or "weaponizing." Bioweapons are a focus of many terrorist groups and have been employed in recent terrorist agendas. Examples of these attacks are discussed in latter chapters.

RISK MANAGEMENT AND PRIORITIZATION IN BIOSECURITY

Risk management is not unique to handling infectious or harmful agents. In fact, risk management is an industry unto itself. Examples are found in insurance, healthcare, financial sectors, construction, and other arenas where specific risks may be presented and subsequently mitigated. In most cases, risks are prioritized against the magnitude of negative impact, or quantifiable loss, and the probability of such an adverse event occurring.

Let's return to the example of airport security. All of the procedures, checkpoints, physical elements, and actions taken by security staff are implemented security measures based on the potential risks associated with commercial air travel. Primarily, these security measures are in place to identify and prevent individuals with malicious intent from gaining access to aircraft or luggage. In this matter, associated risks are explosives, personal weapons, or even chemicals that could incapacitate crewmembers. The security plan is designed to mitigate these risks and includes control points such as explosive device detectors, X-ray scanners, and identification checks. This is a simple version of risk assessment.

A similar approach is used for biosecurity. Sections II and VI of the BMBL provide a good foundation for risk assessments and the development of a biosafety and biosecurity plan.[5] However, unlike many other risk assessment procedures that are defined by ISO standards (i.e., ISO 31000), there are no set standards for biological risk assessments and biosecurity plans. To date, only qualitative measures are available.

The BMBL defines a biological risk assessment as "a process used to identify the hazardous characteristics of a known infectious or potentially infectious agent or material, the activities that can result in a person's exposure to an agent, the likelihood that such exposure will cause a laboratory-associated infection (LAI), and the probable consequences of such an infection."[5] This statement references critical components of a basic risk assessment: identification of the risk, likelihood of the event to occur, and probable outcomes. These three components dictate the foundation of a risk assessment for biological infectious agents.

The laboratory biosecurity plan should also address issues related to the characteristics of biological agents, the impact of an intentional or unintentional release, the overall value to the research program, and the capabilities and motivation of potential adversaries. This mitigation strategy becomes the foundation of the laboratory biosecurity risk and management plan and can evolve into a functioning laboratory biosecurity program. A laboratory biosecurity plan is based on a systematic approach where assets, threats, and vulnerabilities are identified. The risk assessment and mitigation strategies are developed to protect important assets. Fundamentally, there should be a system in place that adequately protects but does not unduly hinder normal operation of the facility. Further chapters of this book will provide specific details of the components of academic, private, and government biosecurity plans and programs.

Both biorisk assessments and biosecurity plans should be developed in a cooperative manner, incorporating the combined expertise of various contributors, such as engineers, scientists, security staff, regulatory personnel, and administrators. This comprehensive approach allows for the identification of the maximum number of risks and the mitigation of said risks.

BASIC COMPONENTS OF BIOSECURITY

Biosecurity, being derived from "security," encompasses physical elements of security typically thought of as "gates, guards, and guns." These may and often do include policy makers and law-enforcement personnel through interaction with the scientific community. But aside from generic security efforts, such as those at commercial banking institutions and airports, biosecurity draws from unique subsets of varying expertise to form a comprehensive, cohesive concept. If we step back and think about what we are trying to secure (biological and infectious agents), we will realize that other expertise is required (i.e., engineering and science). What truly differentiates biosecurity from other types of security is the asset being protected (i.e., the valuable biological material [VBM], dangerous pathogen). These assets are not visible to the naked eye and can reproduce to a theoretically infinite quantity under optimal culture conditions.

Although not comprehensive, the basic components of biosecurity are:

- Physical (gates, guards, guns; see Chapter 4)
- Operational (standard operating procedures, management practices, institutional policies; see Chapter 5)
- Personnel reliability (see Chapter 6)

- Information security (see Chapter 5)
- Risk assessments (see Chapter 2)
- Threat assessments (see Chapter 6)
- Vulnerability assessments (see Chapter 6)

It is also of note that one of the seminal documents in biosafety, BMBL, now includes a section on biosecurity. The 4th edition of the BMBL was the first revision of this document to provide the groundwork of a biosecurity discussion and was updated in the 5th edition in 2007. The BMBL defines biosecurity through a system of risk management: it establishes which, if any, agents require biosecurity measures to prevent loss, theft, diversion, or intentional misuse and ensures that the protective measures provided, and the costs associated with that protection, are proportional to the risk."[5]

CONCLUSION

The previous discussions illustrate the complex issues regarding life-science work as it relates to biorisk, biosafety, and biosecurity. A continuing challenge is the need to expand the basic tenets of laboratory security employed by research laboratories into industry sectors that may not have considered biosecurity a need previously, such as agriculture, mass transportation systems, and public health: all have vulnerabilities to an adverse event of biological origin. But the threat of biological attacks, the multiple locations that infectious agents can be and are located, and the lack of industry standards all point to the fact that institutions need to pay attention to the risks and the opportunity to address laboratory biosecurity and overall biorisk management.

Another challenge is the fragmented nature in which laboratory biosecurity expertise exists. Very few unified guidelines for biosecurity exist. For example, the BMBL[5] and WHO Laboratory Biosafety Manual[3] remain common codes of practice for the biosafety community. No comprehensive guidance document of this nature exists for biosecurity.

This chapter has also touched on the fact that a balance between security policy and the vital role of disease research must be struck. The importance of infectious disease research is clear, with the goal of the development and implementation of new therapeutics that will treat or cure human, animal, and plant disease. Security policy must protect these necessary research endeavors without providing undue hindrance.

Research, clinical, and diagnostic operations utilizing infectious biological agents have increased worldwide. The expansion of laboratories and programs has in some instances outpaced the development of relevant policy. This expansion of capacity has also created a newer challenge for those assigned with laboratory security responsibilities.

Many factors, including the access to detailed information via Internet resources, have made available to the global public specific information regarding the isolation, purification, and production of biological agents. There appears to be a general increase in knowledge that was once isolated to highly educated scientists. As will be discussed in detail in this book, this spread of information potentially sets the stage for access to potentially harmful biological, infectious agents by potential enemies. In short, those individuals who would like to develop weapons derived from biological agents have much more information at their disposal than they did just a few years ago. The ethics and potential impacts of this fact raise alarm.

Biosecurity is a complex package of ideas, practices, policy, and challenges. This book aims to address the major areas of biosecurity and to present a comprehensive analysis of biosecurity as research, health, and policy continues to rapidly and progressively move forward.

References

1. Food and Agriculture Organization of the United Nations. Biosecurity for Agriculture and Food Production. Accessed at http://www.fao.org/biosecurity/.
2. L.A. Meyerson and J.K. Reaser. A unified definition of biosecurity. Science 295: 44 (2002).
3. World Health Organization. Laboratory Biosafety Manual, 3rd edition. Geneva, Switzerland (2004).
4. P.M. Parker, Ph.D. (ed.). Biosecurity: Webster's Timeline History 1928–2007. Icon Group International (2010).
5. U.S. Department of Health and Human Services, Public Health Service, Centers for Disease Control and Prevention, National Institutes of Health. Biosafety in Microbiological and Biomedical Laboratories, 5th edition. In: Chosewood L, Wilson D. (eds.). U.S. Government Printing Office, Washington, DC (2009).
6. World Health Organization. Biorisk Management: Laboratory Biosecurity Guidance (2006).
7. Classification of etiological agents on the basis of hazard. 4th ed. Atlanta, Centers for Disease Control (1974).
8. London School of Tropical Medicine. Report of the Committee of Inquiry into the Smallpox Outbreak in London in March and April 1973. London: H.M.S.O. (1974).
9. 41 Fed Register 27902 (1974).
10. Potential biohazards of recombinant DNA molecules. Proc. Nat. Acad. Sci. 71(7): 2593–2594.
11. 42 CFR Part 73. The Possession, Use, and Transfer of Select Agents and Toxins (2005).

12. U.S. Army Regulation 50-1. Nuclear and Chemical Weapons and Materiel. Biological Surety (2008).
13. NRC Study: Responsible Research with Biological Select Agents and Toxins (2009).
14. C. Skvorc and D.E. Wilson. Developing a Behavioral Health Screening Program for BSL-4 Laboratory Workers at the National Institutes of Health. Biosecur. Bioterror. 9(1): 23–29 (2011 March).
15. Brigadier General Russ Zajtchuk (ed) and Ronald F. Bellamy, M.D. (ed). Medical Aspects of Biological Warfare. Office of the Surgeon General, Department of the Army. 2007.
16. Radovanović Z, Djordjević Z. Mass vaccination against smallpox and mortality in Yugoslavia in 1972. Trans R Soc Trop Med Hyg. 73(1):122 (1979).
17. K. Alibek (with S. Hendelman). Biohazard: The Chilling True Story of the Largest Covert Biological Weapons Program in the World—Told from Inside by the Man Who Ran It. Dell Publishing (1999).
18. Carus, S. Bioterrorism and Biocrimes: The Illicit Use of Biological Agents in the 20th Century. Center for Counterproliferation Research, National Defense University (August 1998, July 1999 revision).
19. United Nations General Assembly. Measures to eliminate international terrorism. A/RES/50/53. 87th plenary meeting (11 December 1995).
20. J.M. Smith and W.C. Thomas (eds.). The Terrorism Threat and U.S. Government Response: Operational and Organizational Factors. U.S. Air Force Institute for National Security Studies. U.S. Air Force Academy, Colorado (2001).

Biosecurity as a Function of Biosafety Microbiological Laboratories

Jonathan Y. Richmond, Ryan N. Burnette, and Gigi Kwik Gronvall

INTRODUCTION

This chapter explains how a comprehensive biosafety program lays the groundwork for enhanced biosecurity within research institutions. Even in nonresearch arenas, where formal biosafety programs are rare, a biosafety program could be useful in providing not only a certain level of biosafety but a great deal of biosecurity. For example, shipping or trucking industries ship biological agents from port to market and therefore could use formal programs to bring both biosafety and biosecurity into their practices. As the legitimate use of biological agents expands, so will the types of industries handling them and their need for biosafety and biosecurity best practices. The implications of biosafety and biosecurity in these "related" industries will be discussed in subsequent chapters.

Biosecurity was first introduced to research laboratories when it became clear that enhanced security, and not just biological safety (biosafety), was required to study infectious disease. In 1995, Larry Wayne Harris, a member of the white supremacist group Aryan Nations, obtained *Yersinia pestis* (the causative agent of plague) from a biological services company. Partly due to the difficulty of prosecuting Harris for obtaining a potential biological weapon, Congress passed the Antiterrorism and Effective Death Penalty Act of 1996.[1] This requires research institutions wishing to possess certain "select" microbiological agents (known as "select agents" and currently called "biological select agents and toxins" or "BSATs") to be registered with the Centers for Disease Control and Prevention (CDC) and follow rigorous transfer procedures. The CDC, in turn, was lawfully required to ensure that the receiving laboratory possessed and practiced sufficient biosafety to handle such agents.

In 1996, there were no specific biosecurity guidelines in microbiology laboratories. Simply maintaining appropriate biosafety levels and biocontainment was sufficient. Best practices for handling select agents are described in detail in the CDC/National Institutes of Health (NIH) publication Biosafety in Microbiological and Biomedical Laboratories, which is commonly referred to as the BMBL,[2] as described in Chapter 1.

EVOLVING STEPS TOWARDS BIOSECURITY

When the 4th edition of the BMBL was published in 1999, it included the first federal requirements for biosecurity as it related to microbiology laboratories.[2] Seven steps to achieve biosecurity were outlined in Appendix F, "Laboratory Security and Emergency Response for Microbiological and Biomedical Laboratories":

1. Recognize that laboratory security is related to, but different than, laboratory safety. Traditional biosafety guidelines emphasize the need for optimal work practices, appropriate equipment for containment of infectious or toxic materials, well-designed facilities, and administrative controls to minimize risk of worker injury or laboratory contamination (versus measures to minimize the potential for *intentional* misuse). This perceived expansion of risks from the accidental to the malevolent is important for assessing risks and threats.
2. Control access to areas where biologic agents or toxins are used and stored. For example, the areas where people work with these agents should be consolidated to the greatest extent, so as to implement security measures more effectively, and should be separated from the public areas of buildings.
3. Know which personnel are in the laboratory area and have reason to access it.
4. Know which materials are brought into the laboratory area.
5. Know which materials are removed from the laboratory area.
6. Have an emergency response plan. Limiting access to the laboratory and animal areas, which is a "best practice," may make emergency response to other areas of laboratory procedures more difficult. This reality needs to be considered as emergency plans are developed and needs to be integrated with institution-wide plans. Plans for emergencies should address a range of adverse events, including bomb threats, severe weather (i.e., hurricanes

or floods), earthquakes, power outages, and other natural or man-made disasters. The emergency plan should address relocation of select agents. Most importantly, the emergency response plan should be exercised with all employees.

7. Have a protocol for reporting incidents of theft, loss, or release.

While the select agent program introduced these "commonsense" security and safety accountability measures, another security gap was exposed following the U.S. anthrax mail incidents of 2001 (see Chapter 3). While the CDC tracked anthrax shipments from one laboratory to another, there was no official accounting system to identify which laboratories possessed anthrax cultures.[3] Thus, in 2001, the U.S. Patriot Act[4] introduced a mechanism for laboratory accountability, as well as many other security provisions that ultimately changed the nature of biological select agents and toxins (BSAT) research.

The new laws prohibited certain restricted persons from shipping, processing, or receiving select agents. What were only *guidelines* in the BMBL became *requirements* for those possessing BSATs, and biosecurity principles were emphasized during facility audits. The CDC was charged with the responsibility to monitor the movement of these agents, ensuring that select agent facilities met BMBL requirements. In 2002, the CDC encouraged laboratory workers to focus on the risk assessment process when planning to work with these agents. They initiated site-specific reviews of physical security; security of data and electronic technology systems; employee security; access controls to laboratory and animal areas, procedures for agent inventory and accountability, shipping/transfer and receiving of select agents, unintentional incident and injury policies, emergency response plans, and policies that address breaches in security.[5]

The Public Health Security and Bioterrorism Preparedness and Response Act of 2002 expanded the responsibility to the Animal Plant Health Inspection Service (APHIS) within the U.S. Department of Agriculture (USDA) to encompass high-consequence animal pathogens and toxins, including zoonotic (or "crossover") select agents.[6] CDC and APHIS now require all personnel interacting with select agents and toxins to undergo a security risk assessment (SRA). The SRA is an electronic records check performed by the Federal Bureau of Investigation's (FBI) Criminal Justice Information Services Division (CJIS). This check determines if personnel meet one of the statutory restrictors, which would either prohibit registration or restrict access, respectively. The results of an SRA will assist APHIS and CDC in its determination that the entity may possess, use, or transfer select agents or toxins or that an individual may have access to regulated biological select agents and toxins (BSAT).

Using the SRA process, the FBI identifies whether an individual or an entity is a restricted person as defined by section 175b of Title 18, United States Code.[7] The FBI uses the SRA process to determine whether an individual has committed a crime set forth in section 2332b(g)(5) of Title 18, United States Code, is knowingly involved with any organization engaging in domestic or international terrorism, as defined in section 2331 of Title 18, United States Code, involved with organizations engaging in intentional acts of violence, or is an agent of a foreign power. as defined in section 1801 of Title 50, United States Code.

The term "restricted person" is an individual who meets the following criteria:[7]

- Is under indictment for a crime punishable by imprisonment for a term exceeding 1 year
- Has been convicted in any court of a crime punishable by imprisonment for a term exceeding 1 year
- Is a fugitive from justice
- Is an unlawful user of any controlled substance (as defined in section 102 of the Controlled Substances Act (21 U.S.C. 802)
- Is an alien illegally or unlawfully in the United States
- Has been adjudicated as a mental defective or has been committed to any mental institution
- Is an alien (other than an alien lawfully admitted for permanent residence) who is a national of a country as to which the Secretary of State, pursuant to section 6(j) of the Export Administration Act of 1979 (50 U.S.C. App. 2405(j)), section 620A of chapter 1 of part M of the Foreign Assistance Act of 1961 (22 U.S.C. 2371), or section 40(d) of chapter 3 of the Arms Export Control Act (22 U.S.C. 2780(d)), has made a determination (that remains in effect) that such country has repeatedly provided support for acts of international terrorism
- Has been discharged from the Armed Services of the United States under dishonorable conditions

BIOSECURITY IS EMBEDDED IN BIOSAFETY PROGRAMS

Laboratories following the latest BMBL[8] guidelines, the WHO Laboratory Biosafety Manual,[9] or similar national biosafety guidelines already have many good laboratory practices in place that provide a strong biosecurity profile. Biosafety and biosecurity are related activities that rely heavily on management and individual acceptance of responsibility for the safe handling of all microbial agents and toxins.

The objective of biosafety is to reduce or eliminate accidental exposure to or the release of potentially hazardous biological agents. This is achieved through various degrees of laboratory containment as well as adherence to best practices for managing infectious materials in a laboratory setting. To understand the risks and the safe practices that must be adhered to, all personnel handling infectious materials should perform a risk assessment, a deliberate process of understanding the risks associated with manipulating microorganisms or undertaking experiments with toxins. The current BMBL provides excellent step-by-step guidance for performing a risk assessment.[8]

The objective of biosecurity is to protect against theft or diversion of high-consequence pathogens and toxins (i.e., dangerous biological agents, BSAT, or valuable biological materials [VBM]) by defining the risks: evaluating probabilities and consequences, protecting the defined assets against defined threats, applying a graded protection approach, and affecting research operations only to the level required. This is built upon biosafety, the collection of administrative policies, work practices, facility design practices, and safety equipment used to protect laboratory workers, their co-workers, and the environment from inadvertent transmission of infectious organisms.

There has been considerable discussion in recent years regarding the term "biosecurity," as it has different meanings in various scientific disciplines and in different languages.[10] In the context of this chapter, biosecurity refers to the prevention of high-consequence microbial agents and toxins or critical relevant information from loss, theft, release, or diversion by those who intend to pursue intentional misuse. The World Health Organization (WHO) defines "biorisk" as the probability or chance that a particular adverse event, possibly leading to harm, will occur.[9] WHO places the responsibility for biorisk management on the facility and its manager (director), who is required to "demonstrate that appropriate and valid biorisk reduction (minimization) procedures have been established and are implemented."[9] A biorisk management committee should be established to assist the facility director in identifying, developing, and reaching biorisk management goals.[9]

A robust risk assessment program should include a review of the potential for laboratory-acquired infections (LAIs) and the minimization of such risks by strict biosafety and biosecurity best practices. These procedures should address access control and agent storage/use activities. It is important to distinguish the differences between a "risk assessment" and a "threat assessment" (defined in Chapter 6).

CONTROL ACCESS TO HIGH-CONSEQUENCE PATHOGENS AND TOXINS

There are several reasons to control access to BSAT laboratories beyond the legal requirements. The laboratory may store or work with proprietary cultures, require good manufacturing practices, use radioisotopes, perform work on research animals, or require chain-of-custody practices—all reasons to restrict access. There are numerous strategies available to implement access security at a range of costs, ranging from armed guards to complex features such as swipe cards, keypads, and biometrics to simple measures such as locks on lab doors and freezers. As an example, badging employees with color-coded swipe cards can identify who has access to certain areas of the facility. Visitors also need to be identified and provided with appropriate escorts. Facility issues and experience with biorisk management have been published and should be consulted as resources for facility managers and biosafety officers.[11, 12, 13]

Honest, reliable, and conscientious workers represent the foundation of an effective security program.[5] Importantly, the person in charge of the laboratory needs to know the people who are working there. There are programs at several larger institutions that have formal personnel reliability procedures to assist in these evaluations. But perhaps the most important ingredients for personnel reliability are leadership and strong management.[14]

Evaluations performed within microbiological laboratories have concluded that personnel issues were the most vulnerable part of a biosecurity program and the most difficult to control.[15] Certainly, the U.S. anthrax attack is one glaring example of the lack of personnel responsibility that can occur. Another example is the increasing number of disruptions by animal rights groups, who often infiltrate animal facilities and subsequently "break in."[16]

ACCOUNTABILITY

Maintaining an accurate inventory control is one hallmark of a robust biosafety/biosecurity program, and it is required for select agents. Of course, microbiological agents are unlike chemical or radiological materials, as microbes can reproduce and multiply. It is therefore impossible to account for every single microorganism. In general, accountability is applied to the seed stocks or cultures rather than to every single tube, flask, or plate used in clinical or research studies. However, this practice is criticized. The National Academies Committee on BSAT research recommended that this lax requirement be changed, as it is "both unreliable and counterproductive, yielding a false sense of security."[17]

At this time, however, inventory control is required by the select agent rule, which encompasses inventory maintenance and accountability for "lost" samples. There is also a requirement to notify CDC or USDA/APHIS when the entire BSAT stock is consumed or destroyed, so it can be officially removed from the facility's inventory. Additionally, significant tracking procedures are required when BSAT samples are transferred to another laboratory, either within the same institution or an outside one.

OCCUPATIONAL HEALTH PROGRAMS

Modern microbiology laboratory managers recognize the need for comprehensive occupational health programs. Typical occupational health programs include employee health evaluations, agent-appropriate immunizations, pulmonary evaluations for persons needing to wear respirators, first-aid training, and occupational exposure evaluations. Some institutions participate in save-serum programs (blood serum can be recovered, saved, and used in other research and treatment applications), but it is necessary to understand the pros and cons of such programs before initiating them.[19]

Occupational health programs can be as simple as having an informed infectious disease physician on call to provide counsel and emergency care up to an in-house clinic fully staffed with appropriate personnel. Many institutions include physicians in the biosafety and biosecurity risk assessment. This is beneficial in the event of employee exposure, particularly when the work involves high-consequence agents that are not normally circulating in the community. When employees participate in fieldwork, it is important not only to equip them with appropriate travel medical kits but also to provide preparedness training for emergency evacuations.

BIOSAFETY CULTURE

Institutions embracing a strong biosafety culture will maintain a biosecurity profile as well.[20] As summarized in the BMBL, the risk assessment and risk management processes for biosafety and biosecurity are interrelated:[8]

- Identify and prioritize biological agents and/or toxins to determine why you have them and what would be the consequence of their misuse
- Identify and prioritize the threat to these biological agents/toxins by either "insiders" or "outsiders" and why your institute might be a target
- Analyze the risk of specific security scenarios that might be employed to secure these agents
- Design and develop an overall risk management program that is commensurate with the potential for intentional or unintentional "release"
- Regularly evaluate your institution's risk posture and protection objectives, following all aspects of receipt, storage, use, and transport of such agents

CONCLUSION

Maintaining high-consequence microorganisms and toxins (i.e., BSATs) in laboratories requires an appreciation of the risks to employee health and the possibility of theft or misuse. Biosafety programs provide a substantial foundation for addressing biosecurity concerns; both risks are manageable in institutions that maintain a strong biosafety culture that encompasses all organizational entities.

References

1. The Antiterrorism and Effective Death Penalty Act 110 Stat. 1214 (1996).
2. U.S. Department of Health and Human Services, Public Health Service, Centers for Disease Control and Prevention, and National Institutes of Health. Biosafety in Microbiological and Biomedical Labroatories, 4th Edition. U.S. Government Printing Office, Washington, DC (1999).
3. Gronvall, G.K. Improving the Select Agent Program. Bulletin of Atomic Scientists Web Edition (October 28, 2008). Available at http://www.thebulletin.org/web-edition/op-eds/improving-the-select-agent-program (accessed July 19, 2012).
4. Uniting and Strengthening America by Providing Appropriate Tools Required to Intercept and Obstruct Terrorism (U.S. Patriot Act) Act of 2001, §Sec. 817. Expansion of the Biological Weapons Statute (2001).
5. Richmond, J. Nesby and O'Dell, S. Laboratory Security and Emergency Guidance for Laboratories Working with Select Agents. Morbidity and Mortality Weekly Report, Centers for Disease Control and Prevention. Atlanta, Georgia. 51(RR-19 [2002]).
6. Public Health Security and Bioterrorism Preparedness and Response Act (2002).
7. National Select Agent Registry, Center for Disease Control and Prevention, U.S. Department of Agriculture. Frequently asked questions concerning Security Risk Assessments (2012). Available at http://www.selectagents.gov/FAQ_SecurityRiskAssessments.html (accessed November 27, 2012).
8. U.S. Department of Health and Human Services, Public Health Service, Centers for Disease Control and Prevention, National Institutes of Health. Biosafety in Microbiological and Biomedical Laboratories, 5th edition. In: Chosewood, L., and Wilson, D. (eds). U.S. Government Printing Office, Washington, DC (2009).

9. World Health Organization. Biorisk management: Laboratory Biosecurity Guidance. WHO/CDS/EPR/2006.6 (2006).
10. Nordmann, B.D. Issues in biosecurity and biosafety. Int J Antimicrob Agents. 36 Suppl 1:S66-69 (November 2010).
11. Richmond, J. and Howard, W. Bio-Risk Assessment in Laboratory Design. In Richmond, J.Y. (ed.). Designing a Modern Microbiological/Biomedical Laboratory: Lab Design Process and Technology. Washington, DC: American Public Health Association (1997, p.55–62).
12. Richmond, J. Biosafety and Biosecurity in Virology: Facility Requirements. Virus Reviews and Research. 10(1):7–13 (2005).
13. Crane, J. and Richmond, J. Design of Biomedical Laboratory Facilities. D.O. Fleming and D.L. Hunt (eds.). Biological Safety, Principles and Practices, 4th edition. Washington, DC: ASM Press (2007).
14. Franz, D.R. and LeDuc, J.W. Commentary: Balancing our approach to the insider threat. Biosecur Bioterror. 9(3):1–2 (2011).
15. Report of the Defense Science Board Task Force on Department of Defense biological safety and security program. Washington, D.C.: Office of the Under Secretary of Defense for Acquisition, Technology, and Logistics (2009).
16. Bailey, Matthew R., Rich, Barbara A., and Bennett, B.T. Crisis Planning to Manage Risks Posed by Animal Rights Extremists. ILAR Journal. 51(2):138–148 (2010). Available at http://dels-old.nas.edu/ilar_n/ilarjournal/51_2/html/pdfs/v5102Bailey.pdf (accessed November 27, 2012).
17. National Research Council (U.S.). Committee on Laboratory Security and Personnel Reliability Assurance Systems for Laboratories Conducting Research on Biological Select Agents and Toxins. National Academies Press (U.S.). Responsible research with biological Select Agents and toxins. Washington, D.C.: National Academies Press (2009). Available at http://site.ebrary.com/lib/pitt/Doc?id=10367629.
18. Richmond, J. Anthology of Biosafety XI. Worker Health and Safety Issues. Mundelein, IL: American Biological Safety Association (2008).
19. Kasting, G. and Hawkins, M. Serum Banking in Biomedical Research. In Richmond, J. (ed.). Anthology of Biosafety, XI. Worker Health and Safety Issues (pages 17–32). Mundelein, IL: American Biological Safety Association (2008).
20. Richmond, J. Achieving a "Biosafety Culture" in Biomedical Laboratories. In Richmond, J. (ed.) Anthology of Biosafety, XI. Worker Health and Safety Issues (pgs. 169–182). Mundelein, IL: American Biological Safety Association (2009).

9. World Health Organization. [illegible]

10. [illegible]

11. [illegible]

12. [illegible]

13. [illegible]

14. [illegible]

15. [illegible]

16. [illegible]

17. [illegible]

18. [illegible]

CHAPTER 3

A Perspective of Biosecurity: Past to Present

Robert J. Hawley and Joseph P. Kozlovac

INTRODUCTION

Although the term biosecurity, or more accurately biological security, has been defined differently in many settings by various professions, it can be considered as a broad set of preventive measures designed to reduce the risk of transmission of infectious agents, toxins, quarantined pests, invasive alien species, and living modified organisms and to implement controls to prevent theft, loss, or diversion of an agent. A more narrowly focused description of biosecurity, adopted from the U.S. Centers for Disease Control and Prevention (CDC), is stated as "the discipline addressing the security of microbiological agents and toxins and the threats posed to human and animal health, the environment, and the economy by deliberate misuse or release."[1] Thus, the goal of this more focused definition is to prevent the intentional theft of biological materials from research laboratories with subsequent potential harm.

Microbiological agents and toxins (biological agents) are either replicating agents (bacteria or viruses) or nonreplicating materials (toxins or physiologically active proteins or peptides) that can be produced by living organisms.[2] The four categories (biological warfare, bioterrorism, biocrimes, and bioaccidents), as described by Katz,[3] succinctly address the potential of biological agents to cause biosecurity incidents. This chapter describes several biosecurity incidents and explains the proactive steps that could have avoided them.

BIOLOGICAL WARFARE

Biological warfare (BW), also known as germ warfare, is the intentional use of microorganisms or toxins derived from living organisms with the intent to kill or incapacitate humans, animals, or plants as an act of war or state-sanctioned political violence, which typically includes military-related activities.[4,5,6] Offensive biological warfare, including the mass production, stockpiling, and use of biological weapons, was outlawed by the 1972 Biological Weapons Convention (BWC). The rationale behind this treaty was to prevent biological attacks, which could conceivably result in large numbers of civilian fatalities and disrupt economic and societal infrastructure. Many countries, including signatories of the BWC, currently pursue research into the defense of or protection against biological warfare, which is not prohibited by the BWC. However, although a signatory to the BWC, the Soviet Union Ministry of Defense (Biopreparat) biological warfare program existed after the country signed the biological weapons convention[2] on April 10, 1972. The extensive program of Biopreparat continued into the early 1990s and included assassinations, accidental contamination of civilian populations, and development of agricultural biowarfare agents. Concerns persist that the program was never terminated.[7]

The intentional use of biological warfare has existed for many centuries and is well documented.[5,6,8-19] Ancient aboriginal people, forefathers of South American tribes, used poison darts to subdue their enemies. Some poisons were obtained from plants, while others came from animals such as the poison arrow frog. In 1346–47 Tatar forces used catapults to hurl plague-infected (*Yersinia pestis*) soldiers into the city to initiate an epidemic during the siege of the Genoese trading post, Caffa.[9,20] The swiftly spreading deaths caused the Genoans to evacuate and sail to Genoa. It is from here that plague is believed to have spread throughout Europe, causing the historical Black Death.[21]

The first recorded use of a biological agent on North American soil (directed at humans) occurred during a conflict known as the Pontiac Rebellion. The Pontiac Rebellion was a war launched in 1763 by a loose confederation of Native American tribes, primarily from the Great Lakes region, the Illinois Country, and Ohio Country. The Native Americans were dissatisfied with British postwar policies after their victory in the French and Indian War (1754–1763). Warriors from numerous tribes joined the uprising in an effort to drive British soldiers and settlers out of the region.[107] The British deliberately used smallpox virus against Native American tribes. The "naïve" Delaware Indians (a tribe of the Ohio Country) were invited to Fort Pitt, where they were given two blankets and a handkerchief contaminated with smallpox virus as gifts.[8,22,23] These "gifts" were a delivery system that the medical world would now refer to as a "fomite," an inanimate object capable of naturally containing or transporting an infectious agent. This method of biological warfare was used again during the American Civil War, when Dr. Luke Blackburn of Kentucky sold clothing contaminated with smallpox and yellow fever to unsuspecting Union troops as a deliberate act to support the Confederacy. Supposedly, at least one Union officer's obituary stated that he died of smallpox. Since more soldiers died of disease during the Civil War than were killed on the battlefield, the effectiveness of Blackburn's work was difficult to assess.[24]

Biological warfare directed against animals occurred during World War I. The Germans attempted to contaminate animal feed and infect livestock for export to Allied forces, with *Bacillus anthracis* (causative agent of anthrax) and *Burkholderia mallei* (causative agent of glanders) and to infect sheep in Romania for export to Russia.[9,25] During 1916–1917, livestock in Argentina destined for export to Allied forces were infected with *B. anthracis* and *B. mallei,* resulting in the death of more than 200 mules.[9,23,26] A German saboteur arrested in Mesopotamia in 1917 had allegedly planned to inoculate mules and infect horses of the French cavalry in France with *B. mallei.*[9,108]

Biological agents can also be used against crops. Since their use requires more effort and technology than in case of anti-animal agents, their use by individuals seems less probable. For example, in case of fungal pathogens, the spores must be obtained, stored, and then disseminated under favorable weather conditions. However, the relatively long period of time between introduction of the agent and discovery of the disease makes the attack difficult to detect. The relatively long period of time between introduction and intended negative impact would make it somewhat undesirable for a fast-paced military action; however, once introduced, it would be hard to detect by the enemy and already widespread at time of detection, which would be a strategic advantage. The pathogens that most likely can be used against crops are *Puccinia graminis, Piricularia orizae,* and *Phytophthora infestans,* causing wheat rust, rice blast, and late blight of potatoes, respectively. Most bacteria pathogenic for plants cause diseases that are of little or no economic importance.[27] Others cause damage that can result in decreased crop yields, but these are usually tolerable. Alternatively, a small number of plant-pathogenic bacteria are capable of causing very serious damage to plants of considerable economic value.[28] The number of bacterial pathogens of plants are too numerous to cite.[29] Epidemic capability in plants depends on several factors: susceptibility of the host, dependence upon weather conditions, such as temperature, humidity, or wind direction, and human activity. The latter can either assist or prevent the outbreak of the disease in crops. The opportunity for disease outbreak is lessened if modern crop protection practices, such as use of resistant varieties and crop rotation, are employed. The possibility of significant losses in crops is high if these practices are not used.[30] During and after World War II, experiments to evaluate the use of some fungal plant pathogens were conducted in several countries, including Germany, Japan, the United Kingdom, the former Soviet Union, and the U.S. Investigations were also conducted on wheat and barley mosaic streak viruses, potato virus, tobacco mosaic virus, and brown grass mosaic virus for use against barley, maize, and thorn apple. Some species of pests were also regarded as possible anticrop agents.[30,31,32] The United States Army stockpiled rice blast (*Pyricularia grisea*), rye stem rust (*P. graminis* f. sp. *secalis*), and wheat stem rust (*Puccinia graminis* Pers. f. sp. *tritici*) for use against crops prior to the 1969 decision to dismantle the U.S. offensive BW program.[8] The former Soviet Union had a substantial program involving agents for use against crops, including wheat rust, rice blast, karnal bunt, and *Fusarium* spp.[33] *Tilletia indica* (causative agent of wheat cover smut), a fungal strain with crop-destroying potential, was evaluated for use as a weapon by Iraq during the 1980s. Research in 1985 at Salman Pak demonstrated that wheat cover smut spores sprayed over immature wheat plants would be lethal to the crop. During 1988, young wheat plants growing in large fields near the town of Mosul were infected with this agent. The infected wheat subsequently was harvested and moved to Fudaliyah for storage. The Iraqis did not attempt to recover fungus from the harvest and said that the infected crop was destroyed in 1990. These findings imply that Iraqi leaders knew that BW were not only antipersonnel weapons, but they could also be used against crops as part of economic warfare.[34] The chronology of some additional accounts of biological warfare is described in Table 3.1.

BIOTERRORISM

As described in Chapter 1, bioterrorism is the threat or use of biological agents by individuals or groups motivated by political, religious, ecological, or other ideological objective.[5,35] These individuals or groups are not otherwise recognized as an extension of the government of a state.[17]

Aum Shinrikyo (Aum Supreme Truth, currently known as Aleph) is a Japanese cult and terrorist organization. The group was founded by Shoko Asahara in 1984. He sought to establish a theocratic state in Japan.[36] In Kameido, Tokyo,

TABLE 3.1 The chronology of some additional accounts of biological warfare[105] (adapted with additions)

Time Period (reference)	Intent or Action	Agent(s) or Action
Japan, Unit 731; Ping Fan, Manchuria, 1932–1945.[22, 94-96]	• Unit 731 (Epidemic Prevention and Water Purification Department of the Kwantung Army) was a covert biological and chemical warfare research and development unit of the Imperial Japanese Army that undertook lethal human experimentation during the Second Sino-Japanese War (1937–1945) and World War II. Shiro Ishii was a physician and Army officer that formed the biowarfare unit outside of Harbin, Manchuria.	• *B. anthracis, Neisseria meningitidis, Shigella* spp., *B. mallei, Salmonella typhosa, Vibrio cholerae, Yersinia pestis,* smallpox virus
	• Chinese cities attacked (cultures tossed into homes and sprayed from aircraft; potentially infected fleas harvested in the laboratory were released from aircraft).	• *Yersinia pestis* and other biological warfare agents
	• Water supplies and food items contaminated.	• *B. anthracis, Shigella* spp., *Salmonella* spp., *V. cholerae*, and *Y. pestis*
World War II; Germany.[97]	Prisoners in Nazi concentration camps forcibly infected with various biological agents.	• *Rickettsia prowazekii, R. mooseri*, hepatitis A virus, and *Plasmodium* spp. • Treated with investigational vaccines and drugs
World War II; Germany, May 1945.[97]	Polluted a large reservoir in northwestern Bohemia.	Sewage
Great Britain; Gruinard Island, 1941–1942.[98]	During the Second World War, trials of *B. anthracis* as a potential agent of biological warfare were carried out on Gruinard Island off the west coast of Scotland.	*B. anthracis*
France; Koch Foundation Laboratories, Paris, 1941.[99]	The French, with German experts, collaborated to develop botulinum toxin in an inert container that could be dropped in bombs from airplanes.	*Clostridium botulinum* toxin
United States, 1943; Camp Detrick, Horn Island, Mississippi, Granite Peak, Utah.[100]	Offensive and defensive biological warfare program. Field-testing was established in Mississippi. In 1944 Dugway Proving Grounds in Utah replaced Mississippi's testing facilities.	*B. anthracis* and *Brucella suis*

(Table continued on next page)

TABLE 3.1 *(Continued)*

Time Period (reference)	Intent or Action	Agent(s) or Action
United States, Pine Bluff, Arkansas, 1950–1953.[102,101]	Biological weapons operations were conducted at Pine Bluff Arsenal from 1953 to 1969. The laboratory produced *Brucella suis* (the causative agent of brucellosis, also called undulant fever) and *Francisella tularensis* (tularemia, or rabbit fever). Hardware for antipersonnel biological cluster bombs was delivered to Pine Bluff for filling with *Brucella suis* to support air force requirements. By 1955, the accelerated program was producing stocks of *B suis* and *F. tularensis* as biological warfare agent.	*Francisella tularensis* and *Brucella suis*
United States, 1955; Camp Detrick	• Aerosol studies (1-million-liter spherical aerosolization chamber) and efficacy of vaccines, prophylaxis, and therapies under development; Program known as Operation Whitecoat, the name given to a medical research program carried out by the U.S. Army at Fort Detrick, Maryland during the period 1954–1973. The program involved conducting medical research using volunteer enlisted personnel. The volunteers, all conscientious objectors and many members of the Seventh-day Adventist Church, were apprised of the purpose and goals of each project before providing consent to participate in any project. The purpose of the research was to defend troops and civilians against biological weapons. (http://ww2.dcmilitary.com/dcmilitary_archives/stories/030205/33582-1.shtml.	• *Francisella tularensis* and *Coxiella burnetii*
	• Production and storage procedures and techniques, aerosolization methods, behavior of aerosols over large geographic areas, and effects of solar irradiation and climatic conditions.	• Simulants *Aspergillus fumigatus*, *B. subtilis* var. *globigii* (BG), and *Serratia marcescens*

Japan, members of the group aerosolized a liquid suspension of *Bacillus anthracis*. They intended to cause an epidemic that would trigger a world war and allow Asahara to rule the world. During 1999 to 2001, microbiologic and genetic tests were conducted on a liquid environmental sample originally collected during the 1993 incident. Nonencapsulated isolates of *B. anthracis* were cultured from the liquid. Genetic tests revealed that all isolates were identical to a strain used in Japan to vaccinate animals against anthrax; this finding was consistent with the cult members' testimony about the strain source. A retrospective case-detection survey was conducted in 1999 to identify potential human anthrax cases associated with the incident, but none were found. The use of an attenuated *B. anthracis* strain (without the ability to produce a protective capsule), low spore concentrations, ineffective dispersal, a clogged spray device, and inactivation of the spores by sunlight are all likely contributing factors to the lack of human cases.[37]

The Aum Shinrikyo group gained international notoriety in 1995 when it carried out its Sarin gas attack. On the morning of March 20, 1995, Aum members released Sarin gas in a coordinated attack on five trains in the Tokyo subway system, killing 13 commuters, seriously injuring 54, and affecting 980 more. It is difficult to obtain exact numbers since many victims were reluctant to come forward.[38,39,40,41] However, some estimates claim as many as 5,000 people were injured by the Sarin gas—both physically and psychologically. If Aleph's attack had been more effective, upwards of 100,000 could have been affected. Prosecutors allege that Asahara was apprised about planned police raids on cult facilities by an insider and ordered an attack in central Tokyo to divert attention away from the group. The plan evidently backfired, and the police conducted huge simultaneous raids on cult compounds across the country.[42] On 10 prior occasions between 1993 and 1995, the cult was unsuccessful in its attempts to disperse various quantities of either *B. anthracis* (anthrax) or botulinum toxin in downtown Tokyo.[22] The cult attempted to obtain Ebola virus (a deadly hemorrhagic virus) from Zaire, Africa, in 1993, and in 1994 discussed using the virus as a biological weapon. The cult also cultured and experimented with the agents of anthrax, Q-fever, cholera, and botulinum toxin.[43,44]

The Rajaneeshees, a group originally from India, constructed a 64,000-acre "Buddhafield" in Wasco County, Oregon, in 1981 to celebrate their doctrine of beauty, love, and guiltless sex.[35,45] To gain political control, the group desired to seek a majority of electoral seats. In 1982, the sect took charge of Antelope, a neighboring municipality, after winning the majority of electoral seats and renamed the town Rajneesh. The Rajaneeshees built a 160-room hotel, shopping mall, and airstrip on the land, which violated zoning and legal restrictions. The Rajaneeshees devised a plan to invite 3,000 homeless people from New York to the commune in an attempt to win electoral control of the entire county.[46] In their efforts to expand the commune, using harsh attacks and unspecified threats, the Rajaneeshees contaminated salad bars in 10 restaurants with *Salmonella tyhpimurium* in September 1984.[35] This attack is the first documented case of bioterrorism in the United States during the 21st century. Dozens of individuals around the county developed severe gastroenteritis (intestinal inflammation). By the time the crisis ended, 751 people tested positive for salmonella.[47] The biological outbreak occurred before the age of computers and automated record-keeping, and consequently it was days before authorities contacted the Epidemic Intelligence Service at the CDC, illustrating the importance of analyzing hospital records to identify outbreaks of biological agents. It was more than a year before the authorities realized that the Rajaneeshees event was an act of deliberate bioterrorism.[48] The group had intended to sicken county residents so that the 4,000 commune members could gain a majority in the election.[49] In 1985, a member of the cult confirmed allegations of the deliberate intent of the attack.[6] This case study illustrates the ease with which an individual or group could conduct a biological attack; this group merely spread the microorganism over publicly-available salad bars. It also stresses the need for tools to both detect biological outbreaks and prepare public officials and first responders for incidents of bioterrorism.[50] This case also denotes the willingness of terrorist groups to use agents that were not necessarily considered in the past by various State-sponsored offensive biological weapons programs, which primarily looked at agents that could be readily disseminated via aerosol rather than a food-borne pathogen.

The potential for terrorist attacks against agricultural targets (agroterrorism) may be considered a subset of bioterrorism. It is defined as "the deliberate introduction of an animal or plant disease with the goal of generating fear over the safety of food, causing economic losses and/or undermining social stability."[51] Agroterrorism is a concern for national security, especially after the incidents of September 2001, but it is not new.[52] Chapter 8 will cover agroterrorism in detail.

BIOCRIME

A biocrime is the threat or use of biological agents for individual objectives such as revenge or financial gain and is usually directed at an individual person.[3,53,106] Two of the earliest recorded examples occurred during 600 B.C.E. when the Assyrians poisoned enemy wells with rye ergot. Ergot is a parasitic fungus that produces mycotoxins, which cause hallucinations and gangrene. Solon of Athens used the herb hellebore (skunk cabbage, a toxic plant) to poison the water supply of the city of Krissa during his siege.[54] Hellebore is a very effective purgative narcotic that causes cardiac arrest. The list of alleged or actual biocrimes is quite extensive.[35,55,56] Two examples of biocrimes will be addressed in this chapter.

The Cold War (approximately 1945–1991) is responsible for the first documented modern assassination using a biological agent, ricin, an agent derived from castor oil plants (*Ricinus communis*).[57,58] The KGB (Komitet Gosudarstvennoy Bezopasnosti or Committee for State Security, the national security agency of the Soviet Union from 1954 until 1991) and the Bulgarian secret police executed a flawless and as yet unsolved murder. Yuri Andropov, the chairman of the Soviet KGB, reportedly authorized technical assistance and training for the operation.[59] On September 7, 1978, Georgi Ivanov Markov, a Bulgarian writer and journalist who worked for the British Broadcasting Corporation (BBC) and Radio Free Europe, left home for work. He was inclined to take the Waterloo Bridge bus to BBC headquarters. As Markov neared the queue of people waiting for the bus about 1:30 PM he suddenly felt a stinging pain in the back of his right thigh. He turned and saw a heavyset man in his forties stooping over to pick up a dropped umbrella. The man hurriedly crossed to the other side of the street and got in a taxi, which then drove away. The event is recalled as the "Umbrella Murder," executed by the alleged assassin Francesco Gullino, code name "Piccadilly." Unconcerned, Markov continued to travel to work where he told his colleagues what happened. He showed one BBC friend a red pimple-like swelling on his thigh. That evening Markov developed a high fever, was taken to a London hospital, and treated for nonspecific blood poisoning. He died three days later, on September 11th, 1978, at the age of 49.[60] Due to the circumstances and statements Markov made to doctors expressing the suspicion that he had been poisoned, Scotland Yard ordered a thorough autopsy of Markov's body. The forensic pathologists discovered a spherical metal pellet the size of a pin-head embedded in Markov's calf. Upon autopsy, a tiny pellet was found in the wound in Markov's thigh. The pellet measured 1.52 mm in diameter and was composed of 90 percent platinum and 10 percent iridium. It had two holes with diameters of 0.35 mm drilled through it, producing an X-shaped cavity.[35] Further examination by experts from Porton Down (Salisbury, Wiltshire, UK) showed that the pellet contained traces of toxic ricin. A sugary substance coated the tiny holes, creating a bubble that trapped the ricin inside the cavities. The specially crafted coating was designed to melt at 37° Celsius, the temperature of the human body. As the pellet was shot into Markov, the coating melted and the ricin was free to be absorbed into the bloodstream and kill him. Regardless of whether the doctors treating Markov had known that the poison was ricin, the result would have been the same, as there is no known antidote to ricin.[61] With an average lethal dose of 1/5,000 gram, it remains a potent BW. The worldwide processing of over 100,000,000 metric tons of castor beans results in a 5 percent waste mash. This waste mash is ricin toxin.[62]

The most widely publicized and costly biocrime in the United States was the series of anthrax mail attacks of 2001. Decontamination efforts totaled more than $1 billion.[63] In September 2001, *B. anthracis* (anthrax) spores were sent in about five letters through the mail within the United States Postal Service to several different locations in Florida, New York City, and Washington, D.C. The index case (or primary case, the initial patient in an epidemiological investigation) was diagnosed in Florida in October 2001 and was the first case of inhalational anthrax to occur in the United States since 1978.[64] All of the letters were posted from Trenton, New Jersey, and from these mailings twenty-two cases of anthrax infection were diagnosed, five of which resulted in death of the victim due to inhalation anthrax. At least 17 buildings were confirmed to have been contaminated with *B. anthracis* spores.[65] It was found that the spores in the letters were from the Ames strain (isolated from a natural outbreak in Texas in 1981), with each letter containing approximately 2 grams of powder containing the spores. This powder is reported to contain between 100 billion and 1 trillion spores per gram.[66] In September and October 2001, five letters containing spores were mailed to United States Senators Patrick Leahy and Thomas Daschle and various media organizations. Subsequently, over 10,000 people were tested and deemed at risk for contraction of anthrax disease. In 1981 the Ames strain was sent from Texas to the United States Army Medical Research Institute of Infectious Diseases (USAMRIID). By November 2001, the genetic analysis of the anthrax spores began. Over the next few years assays were conducted on about 1,070 samples, and it was determined that eight different samples from the letters all came from a common source known as RMR-1029, which was only found at USAMRIID: "By 2007, investigators conclusively determined that a single spore batch created and maintained by Dr. Bruce E. Ivins at USAMRIID was the parent material for the letter spores. An intensive investigation of individuals with access to that material ensued. Evidence developed from that investigation established that Dr. Ivins alone, mailed the anthrax letters."[56]

BIOACCIDENTS

A bioaccident refers to a spill, release, illness, or incident involving a biological agent in a research or clinical diagnostic laboratory or pharmaceutical venue. Bioaccidents will hereinafter be referenced as "incidents." At present, there is no U.S. national database, accessible by all agenices, documenting all incidents ("accident reporting") at medical facilities. This is unfortunate, as many lessons could be learned from such incidents, and concern has been expressed.[67] The Trans-Federal Task Force on Optimizing Biosafety and Biocontainment Oversight recommended the establishment of a voluntary, nonpunitive centralized system for incident reporting, analysis, and information sharing across all high (Biosafety Level 3 [BSL-3]) and maximum containment (Biosafety Level 4 [BSL-4]) research laboratories to ensure the protection of sensitive and private information as necessary.[68] The prompt and detailed reporting of significant laboratory incidents involving high or maximum containment research is another component of comprehensive and effective biosafety and biocontainment oversight. Reporting and analyzing laboratory-associated infections (LAIs) and other laboratory accidents at the institution level provide useful information about improving laboratory safety. By understanding the causality as it relates to a specific accidental exposure or release, it should be possible to prevent similar incidents in the future. The task force further opined that an analysis of reports of laboratory incidents could help improve laboratory safety and oversight, determine why the incidents occurred and how they could be prevented in the future, provide a resource for generating and sharing lessons learned, and point to the need for new or revised guidelines, practices, or training.[68]

Although there is no centralized reporting system for all laboratory incidents at all medical facilities, those facilities working with BSATs (biological agents and toxins)[69] must develop an incident response plan.[70] The plan must "describe the entity's response procedures for the theft, loss, or release of a select agent or toxin, inventory discrepancies, security breaches (including information systems), severe weather and other natural disasters, workplace violence, bomb threats, suspicious packages, and emergencies such as fire, gas leak, explosion, power outage, etc. The response procedures must account for hazards associated with the select agent and toxin and appropriate actions to contain such select sgent or toxin." This plan must be reviewed annually and revised as necessary after any drill or exercise or any incident. Upon discovery of the theft or loss of a select agent or toxin, an individual or entity must immediately notify the CDC or APHIS (Animal and Plant Health Inspection Service,

U.S. Department of Agriculture) and appropriate Federal, State, or local law-enforcement agencies. Thefts or losses must be reported even if the select agent or toxin is subsequently recovered or the responsible parties are identified.[71] The CDC reported 395 cases of potential release events involving 74 microorganisms and toxins at national laboratories working with select agents and toxins during the period 2003–2009.[72] Although much of this information from the CDC is unpublished and is not available to the research community, some reports were published.[73-81] Table 3.2 lists some salient biosafety and containment lessons that can be gleaned from the information presented in these publications.

Laboratory incidents and LAIs due to exposure to a wide variety of bacteria, viruses, fungi, and parasites have been described since researchers began working with these materials during the mid-1800s.[82,83] Although microbiology addresses the prevention, treatment, and cure of infectious diseases, it can, in certain circumstances, be a potentially dangerous discipline. During the past 150 years, there were many cases of researchers who died because of their professional activity (e.g., nursing, fieldwork, self-inoculation, and laboratory procedures such as mouth pipetting and use of a syringe or needle) or came close to death from infection by agents that were the subject of their research[84]—this was a time when the concepts of safety, cleanliness, contagion, and the transmission of disease from person to person, were in their infancy. Incidents have occurred in microbiological laboratories, clinical laboratories, animal facilities, research and development venues, and production (vaccine) installations. It is often difficult to discern whether the incident was a result of a laboratory mishap or was community-acquired. Historical accounts of incidents occurring since the 1850s have been documented.[84-90, 104,105]

However, there has been a decrease in LAIs over the years. This reflects an improved awareness of the hazards of infectious agents and a greater emphasis on elements of laboratory safety and robust training and use of personal protective equipment. Legislation and guidelines that were introduced over time have probably reduced but not eliminated the risk of occupational exposure to infectious or toxin-containing agents. In addition, there have been improvements in laboratory design and equipment.[91,92] Accidents will happen; microbiology work involving etiologic or zoonotic agents represents a very real occupational hazard to those working in a laboratory. What remains disturbing are unreported incidents. Although the numbers are impossible to discern, we are deprived of a potentially valuable tool about how to conduct safer laboratory operations and continue to provide protection to the environment, plants,

TABLE 3.2 Some salient biosafety and biocontainment lessons

Time Period (reference)	Agent	Activity	Contributing Factors/Findings	Outcome	Lesson(s) Learned
March 2000.[73]	*Burkholderia mallei*	Patient handled laboratory equipment containing live *Burkholderia* strains without wearing gloves.	Patient had insulin-dependent diabetes mellitus.	Patient successfully treated with imipenem and doxycycline.	Laboratory containment methods and microbiologic procedures should be in compliance with recommendations in *Biosafety in Microbiological and Biomedical Laboratories*.[1]
March 2002.[74]	*B. anthracis*	Moving vials containing aliquots of *B. anthracis* isolates from a BSC to a freezer. Tops of vials subsequently found to be contaminated with *B. anthracis*.	Procedure done without gloved hands; patient developed scab on right jaw. No other laboratory workers affected. Patient nor 40 coworkers received the anthrax vaccine.	Patient hospitalized for 4 days and treated with intravenous ciprofloxacin and doxycycline.	Importance of safe laboratory procedures and anthrax immunization for workers routinely handling *B. anthracis* isolates.
September 2004.[76]	*F. tularensis*	Laboratory researchers believed they were working with the Live Vaccine Strain (LVS) of *F. tularensis*.	LVS stock used by the researchers was contaminated with Type A tularemia, a wild-type, virulent form of the organism. Workers did not know the parental lineage of the material they were working with. Use of a colony counter examining open plates of *F. tularensis* cultures outside a biosafety cabinet. Failure to identify work-related illness in laboratory staff.	Three researchers working with tularemia had been ill in 2004, with symptoms consistent with pneumonic tularemia. All three were treated with antibiotics and recovered.	Review of the BSL-2 laboratory where research was conducted and interviews with research personnel revealed inconsistencies in laboratory safety practices. There was wide variability in the use of protective equipment and infection control measures.

(Table continued on next page)

TABLE 3.2 *(Continued)*

Time Period (reference)	Agent	Activity	Contributing Factors/Findings	Outcome	Lesson(s) Learned
February 2006.[77]	*Brucella* spp.	Cleaning of an aerosol chamber where *Brucella* spp. was used; exposed to the agent while cleaning the chamber after experimentation.	Facility not authorized by the CDC to aerosolize *Brucella* spp. Unauthorized access to Select Agents and toxins. Laboratory worker not familiar with the specifics of working with *Brucella*.	CDC issued a cease-and-desist order for all work on *Brucella* within the affected high-containment laboratory, as well as all aerosol work. That order was later expanded to include all work with Select Agents and toxins at the facility.	Laboratory containment methods and microbiologic procedures should be in compliance with recommendations in *Biosafety in Microbiological and Biomedical Laboratories*.[1]
March 2006.[77,103]	*Coxiella burnetii*	Researcher performing restricted experiments (aerosol work) with *C. burnetii*.	Several laboratory workers were found to have elevated titers for *C. burnetii*. BSC integrity compromised. Protective clothing and equipment not used. Biosafety Manual did not contain site-specific or agent-specific information. Workers unaware of potential biological hazards. No effective medical surveillance program.	All research on dangerous pathogens known as Select Agents suspended.	Laboratory containment methods and microbiologic procedures should be in compliance with recommendations in *Biosafety in Microbiological and Biomedical Laboratories*.[1]

and animals and therefore potentially remain vulnerable. An editorial stated, "The solution to this problem, both in the United States and in other nations, would be the establishment of a nationwide, voluntary, incident-reporting system coupled with a centralized, integrated mechanism for analyzing incidents and sharing information and lessons learned from both mandatory reports by a nation's regulatory authorities and the voluntary reporting system."[93]

CONCLUSION

Biosecurity is considered a set of preventive measures designed to reduce the risk of transmission of replicating infectious agents (bacteria or viruses), toxins or physiologically active proteins or peptides, quarantined pests, invasive alien species, and living modified organisms and to implement controls to prevent theft, loss, or diversion of an agent. Biological warfare, bioterrorism, biocrimes, and bioaccidents describe potential avenues to create breaches in biosecurity. However, in most cases (as described in this chapter) proactive and prescriptive measures can prevent these threats. The future of biosecurity may reside in a closer interaction with the principles of biosafety, whereby we can learn from behavior and laboratory mishaps. It is without question that sharing lessons learned is a valuable and powerful mechanism to mitigate future incidents.

References

1. Wilson, D.E, and Chosewood, L.C.. Introduction (p. 6). In Wilson and Chosewood (eds.), Biosafety in Microbiological and Biomedical Laboratories. 5th ed. HHS Publication No. (CDC) 21-1112. U.S. Government Printing Office, Washington, D.C. (2009).
2. Sidell, F.R., and Franz, D.R. Overview: Defense Against the Effects of Chemical and Biological Warfare Agents (pp. 4–5). In Sidell et al., (eds.). Medical Aspects of Chemical and Biological Warfare. Office of the Surgeon General, Department of the Army. Washington, D.C. (2009).
3. Katz, R. Biological Weapons: A National Security Problem that Requires a Public Health Response. Working Paper 2001-04. Princeton University, Office of Population Research (2001).
4. Stern, J. Definitions (p. 20). In J. Stern (ed.). The Ultimate Terrorists Harvard University Press, Cambridge, MA (1999a).
5. Dudley, J.P., and Woodford, M.H. Bioweapons, bioterrorism and biodiversity: potential impacts of biological weapons attacks on agriculture and biological diversity. Rev. sci. tech. Off. Int. Epiz. 21(1):125–137 (2002a).
6. Riedel, S. Biological warfare and bioterrorism: a historical review. Baylor University Medical Center Proceedings 17:400–406 (2004).
7. Parker, H.S. Agricultural Bioterrorism: A Federal Strategy to Meet the Threat. McNair Paper 65, Institute for National Strategic Studies, National Defense University (2002). Accessed at http://www.au.af.mil/au/awc/awcgate/ndu/mcnair65.pdf.

8. Christopher, G. W, Cieslak, T.G., Pavlin, J. A., and Eitzen, E. M. Biological warfare: a historical perspective. J. Am. Med. Assoc. 278:412:417 (1997).
9. Eitzen, E. M., and Takafuji, E. T. Historical overview of biological warfare (pp. 415–423). In R. Zajtchuk and R. F. Bellamy (eds.). Medical Aspects of Chemical and Biological Warfare. Washington, DC: Borden Institute (1997).
10. Hawley R. J., and Eitzen, E. E. Bioterrorism and Biological Safety. In Fleming, D. O., and Hunt, D. L. (eds.). Biological Safety Principles and Practices, 3rd ed., ASM, Washington, DC (2000).
11. Hawley, R. J., and Eitzen, E. E. Protection Against Biological Warfare Agents. In Block, S. (ed.). Disinfection, Sterilization, and Preservation, 5th ed, Lippincott, Williams & Wilkins, Philadelphia, PA (2001a).
12. Hawley, R.J., and Eitzen, E.M. Biological Weapons: A Primer for Microbiologists. Ann. Rev. Microbiol. 55:235–53 (2001b).
13. Katz, R. 2001. Biological Weapons: A National Security Problem that Requires a Public Health Response. Working Paper 2001-04. Princeton University, Office of Population Research.
14. Metcalfe, N. A short history of biological warfare. Medicine, Conflict and Survival 18 (3): 271–282 (2002).
15. Frischknecht, F. The History of Biological Warfare. Human experimentation, modern nightmares and lone madmen in the twentieth century. EMBO reports 4: S47–S52 (2003).
16. Johnson, T.J. A History of Biological Warfare from 300 B.C.E. to the Present. American Association for Respiratory Care (2004). Available from http://www.aarc.org/resources/biological/history.asp.
17. Martin, J.W., Christopher, G.W., and Eitzen, E.E. History of Biological Weapons: From Poisoned Darts to Intentional Epidemics. In D.E. Lounsbury and R.F. Bellamy (eds.), Textbook of Military Medicine. Borden Institute, Washington, DC (2007).
18. Frischknecht, F. The History of Biological Warfare. In Richardt, A., and Blum, Marc-Michael (eds.). Decontamination of Warfare Agents. Wiley, VCH, Verlag GmbH & Co. KGaA. Weinheim (2008).
19. Dando, M.R. and Nixdorff, K. An Introduction to Biological Weapons (pp. 1–12). In McLaughlin, K., andNixdorff, K. (eds.). BWPP (BioWeapons Prevention Project) Biological Weapons Reader. Geneva (2009). Accessed at http://www.bwpp.org/documents/BWPP%20BW%20Reader_final+.pdf.
20. Wheelis, M. Biological Warfare at the 1346 Siege of Caffa. Emerg. Infect. Dis. 8(9):971–975 (2002).
21. Derbes, V.J. De Mussis and the great plague of 1348: a forgotten episode of bacteriological war. J. Amer. Med. Assoc. 196:59–62 (1966).
22. Langford, R.E. Brief History of Biological Weapons. In Langford, R.E. (ed.). Introduction to Weapons of Mass Destruction. John Wiley & Sons, Inc.,Hoboken, NJ (2004).
23. Weller, R.E. Risk of Disease Spread Through Bioterrorism. Veternaria Italiana 42(4):351–367 (2006).
24. Smart, J.K. History of Chemical and Biological Warfare: An American Perspective. In Zajtchuk, R., and Bellamy, R.F. (eds.). Textbook of Military Medicine, Borden Institute. Washington, DC (1997).
25. Wheelis, M. 1999. Biological sabotage in World War I (pp. 35–62). *In* Geissler, E., and van Courtland Moon, J.E. (eds.). Biological and Toxic Weapons: Research

Development, and Use from the Middle Ages to 1945. International and Peace Research Institute. Stockholm, Sweden (1999).

26. Carus, W. S. Bioterrorism and Biocrimes: the Illicit Use of Biological Agents in the 20th Century. Center for Counterproliferation Research. National Defense University, Washington, D.C (1998b).

27. Bradbury, J. F. Guide to Plant Pathogenic Bacteria. CAB International, Kew, UK (1986).

28. Vidhyasekaran, P. Bacterial disease resistance in plants. Molecular biology and biotechnological applications. The Haworth Press, Binghamton, NY (2002).

29. Bull, C.T., De Boer, S.H., Denny, T.P., Firrao, G., Fischer-Le Saux, M., Saddler, G.S., Scortichini, M., Stead, D.E., and Takikawa, Y. Comprehensive List of Names of Plant Pathogenic Bacteria, 1980–2007. J. Plant Pathol. 92 (3):551–592 (2010).

30. Whitby, S.M. The potential use of plant pathogens against crops. Microbes Infect. 3, 73–80 (2001).

31. Dudley, J.P., and Woodford, M.H. Bioweapons, biodiversity, and ecocide: potential effects of biological weapons on biological diversity. BioScience 52:583–592 (2002b).

32. Madden, L.V., and Wheelis, M. The threat of plant pathogens as weapons against U.S. crops. Ann. Rev. Phytopathol. 41:155–176 (2003).

33. Wilson, T.M., Logan-Henfrey, L.,Weller, R.,and Kellman, B. Agroterrorism, Biological Crimes, and Biological Warfare Targeting Animal Agriculture (pp. 23-58). In Brown,C., and Bolin, C. (eds.) Emerging Diseases of Animals. ASM Press. Washington, D.C (2000).

34. Zilinskas, R.A. Iraq's Biological Weapons. The Past as Future? J. Amer. Med. Assoc. 278(5):418–424 (1997).

35. Carus, W.S. *Bioterrorism and Biocrimes. The Illicit Use of Biological Agents Since 1900.* Working Paper. Center for Counterproliferation Research, National Defense University. Washington, D.C. (2001).

36. Stern, J. Getting and Using the Weapons (pp. 60–65). In Stern, J. (ed.). The Ultimate Terrorists. Harvard University Press. Cambridge, MA. (1999b).

37. Takahashi, H., Keim, P., Kaufmann, A. F., Keys, C., Smith, K.L., Taniguchi, K., Inouye, S., and Kurata, T. *Bacillus anthracis* Incident,Kameido, Tokyo, 1993. Emerging Infectious Diseases 10(1):117–120 (2004).

38. Cole, L.A. The Specter of Biological Weapons. Scientific American 12:30–35 (1996).

39. Cole, L.A. Introduction: Testing Ourselves (p. 2). In Cole (ed.). The Eleventh Plague. W.H. Freeman and Company. New York (1997a).

40. Cole L.A. Tools of Terror (pp. 154–156). In Cole (ed.). The Eleventh Plague. W.H. Freeman and Company. New York (1997b).

41. Murakami, H., Birnbaum, A., and Gabriel, P. Underground: The Tokyo Gas Attack and the Japanese Psyche. Random House. New York (2001).

42. Danzig, R., Sageman, M., Leighton, T., Hough, L., Yuki, H., Kotani, R., and Zachary, M. Aum Shinrikyo Insights Into How Terrorists Develop Biological and Chemical Weapons. Center for a New American Security. Washington, DC (2011).

43. Olson, K.B. Aum Shinrikyo: Once and future threat? Emerg. Infect. Dis. 5:513–516 (1999).

44. Tucker, J.B. Historical trends related to bioterrorism: An empirical analysis. Emerg. Infect. Dis. 5:498–504 (1999).

45. Miller, J., Engelberg, S., and Broad, W. The Attack (pp.15-16). In Miller, Engelberg, and Broad (eds.). Germs: Biological Weapons and America's Secret War. Simon & Schuster. New York (2001a).
46. Miller, J., Engelberg, S., and Broad, W. The Attack (p.17). In Miller, Engelberg, and Broad (eds.). Germs: Biological Weapons and America's Secret War. Simon & Schuster. New York (2001b).
47. Miller, J., Engelberg, S., and Broad, W. The Attack (p.19). In Miller, Engelberg, and Broad (eds.). Germs: Biological Weapons and America's Secret War. Simon & Schuster. New York (2001c).
48. Miller, J., Engelberg, S., and Broad, W. The Attack (p.23). In Miller, Engelberg, and Broad (eds.). Germs: Biological Weapons and America's Secret War. Simon & Schuster. New York (2001d).
49. Miller, J., Engelberg, S., and Broad, W. The Attack (p.28). In Miller, Engelberg, and Broad (eds.). Germs: Biological Weapons and America's Secret War. Simon & Schuster. New York (2001e).
50. Török, T.J., Tauxe, R.V. , Wise, R.P., Livengood, J.R., Sokolow, R., Mauvais, S, Birkness, K.A., Skeels, M.R., Horan, J.M., and Foster, L.R. A Large Community Outbreak of Salmonellosis Caused by Intentional Contamination of Restaurant Salad Bars. J. Amer. Med. Assoc. 278(5):389-395 (1997).
51. Monke, J. Congressional Research Service Report for Congress. Agroterrorism: Threat and Preparedness. Agriculture as a Target of Terrorism (2007). Accessed at http://www.fas.org/sgp/crs/terror/RL32521.pdf.
52. Olson, D. Agroterrorism: Threats to America's Economy and Food Supply (2012). Accessed at http://www.fbi.gov/stats-services/publications/law-enforcement-bulletin/february-2012/agroterrorism.
53. Mangold, T., and Goldberg, J. Truth and reconciliation (pp. 261–262). In Mangold and Goldberg (eds.). Plague Wars: A True Story of Biological Warfare. Macmillan Publishers Ltd. London, UK (1999b).
54. Woods, J.B. History of Biological Warfare and Current Threat (p. 2). In Woods (ed.). Medical Management of Biological Casualties Handbook, 6th ed. U.S. Army Medical Research Institute of Infectious Diseases. Fort Detrick, MD (2006)
55. Schutzer, S.E., Budowle, B., and Atlas, R. M. Biocrimes, Microbial Forensics, and the Physician. PLoS Medicine 2(12): 1242–1247 (2005).
56. Federal Bureau of Investigation. Amerithrax Investigative Summary. United States Department of Justice. Washington, DC (2010). Available from: http://www.justice.gov/amerithrax/docs/amx-investigative-summary.pdf.
57. Crompton, R., and Gall, D. Georgi Markov—death in a pellet. Med Leg J. 48:51–62 (1980).
58. Roxas-Duncan, V.I. and Smith, L.A. Of Beans and Beads: Ricin and Abrin in Bioterrorism and Biocrime. J Bioterr Biodef S7:002 (2012).
59. Puddington, A. Bombs, Spies, Poisoned Umbrellas (pp. 244–245). In Puddington (ed.). Broadcasting Freedom: The Cold War Triumph of Radio Free Europe and Radio Liberty. University Press of Kentucky. Lexington, KY (2000).
60. Poli, M., Roy, Chuebner, K.D., Franz, D.R., and Jaax, N. Ricin (pp. 323–335). In Dembek, Z.F. (ed.). Medical Aspects of Chemical and Biological Warfare. Office of the Surgeon General, Department of the Army. Washington, D.C. (2007).
61. Audi, J., Belson, M., Patel, M., Schier, J., and Osterloh, J. Ricin Poisoning A Comprehensive Review. J. Am. Med. Assoc. 294(18):2342–2351 (2005).

62. Shea, D.A. and Gottron, F. Ricin: Technical Background and Potential Role in Terrorism. Congressional Research Service 7-5700: RS21383 (2010).
63. Cordesman, A.H., and Burke, A.A. The Challenge of Biological Terrorism: When to "Cry Wolf," What to Cry, and How to Cry It (p. 3). Center for Strategic and International Studies. Washington, DC (2005). Accessed at: http://www.dsjf.org/Bio_Files/Challenge%20of%20Bioterrorism%20-%20May%2005.pdf.
64. Bush, L.M., Abrams, B.H., Beall, A., and Johnson, C.C. Index Case of Fatal Inhalation Anthrax Due to Bioterrorism in the United States. N. Engl. J. Med. 345(22):1607–1610 (2001).
65. Herrmann, J.G. Advancing National Security Through Science and Post *Amerithrax*: Advancing the Science and Engineering of Decontamination. Science Matters 1(8): 3–7 (2011).
66. Inglesby, T.V., O'Toole, T., Henderson, D.A., Bartlett, J.G., Ascher, M.S., Eitzen, E., Friedlander, A., Gerberding, J., Hauer, J., Hughes, J., McDade, J., Osterholm, M.T., Parker, G., Perl, T.M., Russell, P.K., and Tonat, K. Anthrax as a Biological Weapon. Updated Recommendations for Management. J. Amer. Med. Assoc. 287(17):2236–2253 (2002).
67. Bavoli, P.M. Federal Indifference to Laboratory-Acquired Infections. Appl. Biosafety 10(1):5–7 (2005).
68. Report of the Trans-Federal Task Force on Optimizing Biosafety and Biocontainment Oversight (p. 182 [2009]). Accessed at: http://www.ars.usda.gov/is/br/bbotaskforce/biosafety-FINAL-REPORT-092009.pdf.
69. Code of Federal Regulations. Title 42 Part 73. Select Agents and Toxins. Part 73.3–73.4. HHS Select Agents and toxins and Overlap Select Agents and Toxins. Government Printing Office Washington, DC (2012). Accessed at: http://ecfr.gpoaccess.gov/cgi/t/text/text-idx?c=ecfr&sid=0bdccb0b01af8755841299bd71c6a453&rgn=div8&view=text&node=42:1.0.1.6.61.0.19.3&idno=42.
70. Code of Federal Regulations. Title 42 Part 73. Select Agents and Toxins. Part 73.1., Incident Response. Government Printing Office. Washington, DC (2011a).
71. Code of Federal Regulations. Title 42 Part 73. Select Agents and Toxins. Part 73.19. Notification of theft, loss, or release. Government Printing Office. Washington, DC (2011b).
72. NRC. Protecting the Frontline in Biodefense Research: The Special Immunizations Program. National Research Council of the National Academies. The National Academies Press. Washington, DC (2011). Accessed at :http://www.nap.edu/catalog.php?record_id=13112.
73. CDC. Laboratory-acquired human glanders—Maryland, May 2000. Morbidity and Mortality Weekly Report 49(24):532-535 (2000). Accessed at: http://www.cdc.gov/mmwr/PDF/wk/mm4924.pdf.
74. CDC. Suspected cutaneous anthrax in a laboratory worker—Texas, 2002. Morbidity and Mortality Weekly Report 51(13):279–281 (2002a). Accessed at: http://www.cdc.gov/mmwr/PDF/wk/mm5113.pdf.
75. CDC. Public Health Dispatch: Update: Cutaneous anthrax in a laboratory worker-Texas, 2002. Morbidity and Mortality Weekly Report 51(22):482 (2002b). Accessed at: http://www.cdc.gov/mmwr/PDF/wk/mm5122.pdf.
76. Barry, M.A. Report of Pneumonic Tularemia in Three Boston University Researchers: November 2004–March 2005. Boston Public Health Commission (2005). Accessed at: http://www.cbc.arizona.edu/sites/default/files/Boston_Univerity_Tularemia_report_2005.pdf.

77. U.S. Government Accountability Office. High-Containment Biosafety Laboratories: Preliminary Observations on the Oversight of the Proliferation of BSL-3 and BSL-4 Laboratories in the United States. Testimony Before the Subcommittee on Oversight and Investigations, Committee on Energy and Commerce, House of Representatives. GAO-08-108T. Washington, DC: U.S. Government Accountability Office (2007). Accessed at: http://www.gao.gov/new.items/d08108t.pdf.

78. Unknown author. Accidental Exposure in Biosafety Laboratories (2007). Accessed at: http://fairfoods.org/stopthebiolab/stopthebiolab/documents/accidents-list.pdf.

79. CDC. Laboratory-acquired brucellosis-Indiana and Minnesota, 2006. Morbidity and Mortality Weekly Report 57(02):39–42 (2008). Accessed at: http://www.cdc.gov/mmwr/PDF/wk/mm5702.pdf.

80. National Research Council. Evaluation of the Health and Safety Risks of the New USAMRIID High Containment Facilities at Fort Detrick, MD. National Research Council of the National Academies. The National Academies Press. Washington, DC (2010). Accessed at: http://www.nap.edu/catalog.php?record_id=12871.

81. CDC. Fatal Laboratory-Acquired Infection with an Attenuated *Yersinia pestis* Strain—Chicago, Illinois, 2009. Morbidity and Mortality Weekly Report 60(7):201–205 (2011). Accessed at: http://www.cdc.gov/mmwr/pdf/wk/mm6007.pdf.

82. Reisman, D. Two cases of diphtheria, one a laboratory infection and one in an infant eleven days old. Philadelphia Med. J. 1: 423–426 (1898).

83. Kisskalt, K. Laboratoriumsinfektionen mit Typhusbazillen. Zbl. Hyg. Infekt. 80:145–162 (1915).

84. Piqueras, M. Microbiology: A dangerous profession? Intl. Microbiol. 10:217–226.

85. Sulkin, S.E., and Pike, R.M. Survey of laboratory-acquired infections. Am. J. Pub. Hlth. 41:769–781 (1951).

86. Anderson, R.E., Stein, L., Moss, M., and Gross, N. Potential infectious hazards of common bacteriological Techniques. J. Bacteriol. 64:473–481 (1952).

87. Pike, R.M. Laboratory-associated infections. Summary and analysis of 3921 cases. Health Lab. Sci. 13:105–114 (1976).

88. Pike, R.M. Laboratory-associated infections: incidence, fatalities, causes, and prevention. Ann. Rev. Microbiol. 33: 41–66 (1979).

89. Sewell, D.L. Laboratory-associated infections and biosafety. Clin. Microbiol. Rev.8:389–405 (1995).

90. Rusnak, J.M., Kortepeter, M.G., Hawley, R.J., Anderson, A.O., Boudreau, E., and Eitzen, E. Risk of occupationally acquired illnesses from biological threat agents in unvaccinated laboratory workers. Biosec. and Bioter. 2:1–13 (2004).

91. Crane, J.T., and Richmond, J.Y. Design of Biomedical Laboratory Facilities (p. 280). In Fleming, D.O, and Hunt, D. L. (eds.). Biological Safety Principles and Practices, 4th ed. ASM Washington, DC (2006).

92. Kimman, T.G., Smit, E., and Klein, M. R. Evidence-Based Biosafety: a Review of the Principles and Effectiveness of Microbiological Containment Measures. Clin. Microbiol. Rev. 21(3):403–425 (2008).

93. Kozlovac, J. Laboratory-Acquired Illnesses, Accidents, and Incidents: Turning a Negative into a Positive. Infectious Diseases. Journal of Pakistan. 20(3): 322–324 (2011).

94. Williams, P., and Wallace, D. The Secret of Secrets (pp. 31–50). In Williams, and Wallace (eds.). Unit 731: Japan's Secret Biological Warfare in World War II. MacMillan, Inc. New York (1989a).
95. Harris, S. H. Human experiments: "secrets of secrets" (pp. 59–66). In Harris (ed.) Factories of Death, Japanese Biological Warfare 1932–45 and the American Cover-up. Routledge. New York (1995a).
96. Williams, P., and Wallace, D. Waging germ warfare (pp. 68–70). In Williams and Wallace (eds.). Unit 731: Japan's Secret Biological Warfare in World War II. MacMillan. New York (1989b).
97. Mitscherlich, A., and Mielke, F. Medizin ohne Menschlichkeit: Documente des Nurnberger Arztepoozesses. Frankfurt am Main, Germany. Fischer Taschenbuchverlag (1983).
98. Manchee R.J., and Stewart, R. The decontamination of Gruinard Island. Chem. Br. 24:690–691 (1988).
99. Harris, S. H. 1995b. The United States BW Program (pp. 149-159). In Harris (ed.). Factories of Death, Japanese Biological Warfare 1932–45 and the American Cover-up. Routledge. New York (1995b).
100. Endicott, S., and Hagerman, E. World War II origins (pp. 31–35). In Endicott and Hagerman (eds.). The United States and Biological Warfare: Secrets from the Early Cold War and Korea. Indiana University Press. Bloomington, IN (1998).
101. Mangold, T., and Goldberg, J. Arms race (pp. 34–36). In Mangold, and Goldberg (eds.). Plague Wars: A True Story of Biological Warfare. Macmillan Publishers Ltd. London, UK (1999a).
102. Franz, D.R., Parrott, C., and Takafuji, E.T. 1997. The U.S. Biological Warfare and Biological Defense Programs (pp. 425–436). In Zajtchuk, R., and Bellamy, R.F. (eds.). Textbook of Military Medicine. Borden Institute. Washington, DC (1997).
103. Texas A&M Response to CDC 12-5-07 (2007). Accessed at: http://rgs-portal.tamu.edu/cdc/TexasA_MResponsetoCDC12-5-07.pdf.
104. Harmon, K. Nearly 400 Accidents with Dangerous Pathogens and Biotoxins Reported in U.S. Labs over 7 Years. Sci. Amer. Observations (2011). Accessed at: http://blogs.scientificamerican.com/observations/2011/10/03/nearly-400-accidents-with-dangerous-pathogens-and-bio-toxins-reported-in-u-s-labs-over-seven-years/.
105. Hawley, R.J., and Duley, T.D. Biosafety and Biosecurity: Regulatory Impact) pp. 587–596). In Fleming, D.O., and. Hunt, D.L (eds.). Biological Safety Principles and Practices, 4th ed. ASM. Washington, DC (2006).
106. Mangold, T., and Goldberg, J. 1999. Apocalypse Delayed (pp. 335–351). In Mangold and Goldberg (eds.). Plague Wars: A True Story of Biological Warfare. Macmillan Publishers Ltd. London, UK (1999).
107. Parkman, F. The massacre (pp. 338–367). In F. Parkman (ed.). The conspiracy of Pontiac and the Indian war after the conquest of Canada. 8th ed., Little, Brown, and Company. Boston, MA (1877).
108. Geissler, E. Introduction. In E. Geissler (ed.), Biological and Toxin Weapons Today. Oxford, England. Oxford University Press, Stockholm International Peace Research Institute (1986).

PART II

ELEMENTS OF BIOSECURITY

CHAPTER 4

Physical Elements of Biosecurity

Jeff Slayton, Daniel N. Apple, and Benjamin A. Perman

INTRODUCTION

Biosecurity has many meanings, as evidenced by previous chapters of this book. Biosecurity, as it is used by security professionals responsible for protecting high-containment laboratories, their work, and the personnel vital to the success of those facilities, will be discussed in this chapter. For some, the term "biosecurity" conjures up visions of laboratories with elaborate vacuum chambers, airlocks, and spacesuit-clad researchers. For others, the vision is much more benign—that of the front door of the local veterinarian who treats the animals she works with every day. Others may see biosecurity strictly as a mechanism to protect the integrity of a biological sample and not consider its physical aspects. This chapter provides a "tool kit" to facilitate the protection of biological operations and high-containment laboratories, the biological materials in them, and the researchers themselves. As the debate over what is covered by biosecurity continues, these institutions will need effective protection. This chapter provides a detailed review of the physical measures involved in biosecurity.

PROTECTING AND SECURING BIOLOGICAL MATERIALS

From an academic perspective the physical protection of biological agents is not unlike protecting any other valuable asset. As with other security disciplines, the goal is to prevent the loss or misuse of property and information through theft, sabotage, or unauthorized manipulation. Whether protecting the gold reserves of Fort Knox, the "black gold" of the U.S. Strategic Petroleum Reserve, or a nuclear power plant, success lies in identifying the threat and building an effective mitigation program[1] (see Chapter 6). This is accomplished through an integrated program of security, building design, and operational practices to deter, detect, and neutralize potential threats.

Securing biological material, however, has some characteristics that make it a unique challenge. Unlike many traditional applications of security, the most valuable resource—personnel—is also the greatest potential threat to the institution. Just how much of a danger employees pose is difficult to quantify. One reason for that difficulty is perhaps our greatest source of hope: we have had too few instances of theft, loss, or diversion to truly make robust mathematical models to quantify the risk. The potential for extreme consequences due to the loss or misuse of biological materials forces us to apply our best efforts to securing biological enterprises despite this dearth of evidence.

Understanding exactly what is being secured is of vital importance. It is relatively easy to get lost in this effort. An array of fields can all be analyzed, with each set of assets presenting its own set of challenges. Variations on security abound, depending on the target being protected, the perceived threat, and the building design. Complicating the effort is the temptation to merge biosecurity with other security fields, such as information technology security or nuclear security. To truly understand and apply biosecurity, one must ask several key questions. Why is the security of biological materials different than than that for other materials? What is being protected? What are the best tools to accomplish these tasks?

Biosecurity is Unique

First, the protection of biological agents is unique in that biological agents cannot be observed by the unaided eye. The state of current and immediate future technology does not allow removed or "missing" biological material to be detected. Many applications used in traditional security are therefore ineffective in biosecurity. Unlike other security fields, the capacity to remove biological materials from a facility exceeds the ability to detect their removal. Additionally, the act of removing materials from a facility is unlikely to be observed as an unauthorized act. A further defining feature of biological materials is their nature. Pathogens are living organisms, progressing through a life cycle of replication and expiration. This makes inventory of biological materials an imprecise tool for determining their theft, loss, or diversion.

Second, the life cycle of pathogens increases the consequences of their loss. The loss of a minuscule amount of live pathogens can be used to generate a virtually unlimited supply of the same material, with little to no loss in efficacy of the original pathogen. In the case of biological select agents and toxins (BSATs), the consequences of such a loss are clear and potentially calamitous. By way of example, compare the impact of loss between equally small (microscopic) amounts of the nerve agent Soman (enriched uranium) and the select agent Marburg virus (a hemorrhagic fever virus). The first will be scarcely enough to impact a person exposed to it. The second material, however, is potentially enough to produce a theoretically infinite supply of a pathogen. Note that this example was not chosen by chance: Marburg virus was considered a priority biological weapon for development by the Soviet Union.[2]

In the case of losing nonpathogenic research materials the consequences are much less clear but can be extremely damaging. The theft of intellectual property is particularly damaging to research and development activities,[3] which can be as simple as losing research or as damaging as the loss of competitiveness and viability to the larger institution. Losing the ability to publish a "breakthrough" because of competitor theft can have detrimental career consequences.

Another complicating factor with pathogens, unlike their weapons of mass destruction (WMD) counterparts, is that they exist in nature at varying degrees of utility. One could argue that uranium (a component of many nuclear weapons) extracted from the ground does not provide utility until it is refined. Refining uranium is no small task and more than likely would attract attention from international intelligence agencies. Chemicals, likewise, must be manufactured. Pathogens, however, only need to be recovered and developed into a biological weapon (BW). Fortunately, merely recovering a pathogen does not make it a viable weapon. Arguably, this increases the need for biosecurity. While one can find biological agents in the wild, one faces the challenge of isolating them and safely passaging and replicating them into a viable weapon. This is not the case with biological materials acquired at a laboratory. The materials at a laboratory are separated, identified, and labeled for convenience. Their viability as a weapon is much higher than the same quantity of material recovered in the wild. Therefore, physical security measures at labs containing dangerous biological agents are of vital importance.

Laboratories also provide additional targets. Valuable biological material (VBM), which may not be "dangerous," not only includes pathogenic materials but others that are uniquely biological and valuable. Vaccines and vaccine stocks are among the most obvious targets. Less obvious but equally valuable are laboratory notebooks, results of experiments on magnetic media, sensitive processes, and standard operating procedures (SOPs). The former may be valuable to an individual who seeks to expedite the discovery, passage, and replication process. Where does he need to go to find materials that can be weaponized? He need only read the lab notes of a well-run field collection. How does the scientist replicate the organism without becoming infected? That information is in the lab notes. Which stock of materials is particularly useful to the would-be usurper? That information could have been left in the memory of the last instrument to be used. Combined with the ability to collect viable materials in nature, this makes the need for effective biosecurity self-evident.

In light of all these challenges, one must truly understand what is required by law and regulation and perform a comprehensive risk assessment. These requirements are mandated by Federal Select Agent Regulations (see Chapters 2, 7, and 11) for public, animal, and agricultural health.

Regulations and Physical Biosecurity are Not Prescriptive

Unlike the installation of a basic home alarm system, where the homeowner can choose the level of protection, biological facilities in the U.S. possessing BSATs must comply with federal regulatory requirements.[4,5]

The regulations largely overlap, as they all refer to the possession, use, and transfer of BSATs. The Code of Federal Regulations (CFR) requires each entity using BSAT to register and acquire accreditation from the CDC/HHS. In addition, CDC/HHS uses the Microbiological and Biomedical Laboratories, 5th Edition (BMBL), for its own regulatory compliance.[6] However, these regulations provide no absolute criteria for physical security and are not prescriptive. Although the means to accomplish the task is not prescribed, the requirement is clear: one must be able to demonstrate compliance with the law. This is done to avoid dictating solutions in a rapidly evolving field, where technological solutions of one year ago are often dwarfed by the capabilities of emerging technologies a few years later. Each accreditation source dictates that a security plan must be developed based on a site-specific risk assessment providing graded protection in accordance with the risk of the select agent or toxin. These plans must in turn describe procedures for security to include prevention of loss or theft of BSAT. There are three distinct assessments or "building blocks" that support the development of an effective, compliant security plan. Each assessment complements the other but does not present a complete analysis on its own.

The Risk Assessment

The risk assessment provides insight into which agents or toxins are being worked with, what risks could be posed to the public, and how those agents may be at risk. Risks may include release of pathogens to the public, loss of sensitive research data, or operational disruption. With this information, a facility can ensure that physical security efforts, which can be costly, are directed where needed.

The Threat Assessment

If protection is needed, a threat assessment should be conducted to determine who or what poses a threat, how that threat could manifest, the probability of the threat occurring, and its consequences. A comprehensive threat assessment provides information on what types of person or organization would find the facility an attractive target, what means they could be expected to employ, and the likelihood of occurrence. As an example, a facility that conducts research with live animals may find itself at greater risk of animal-rights activists attempting to disrupt operations than of attack from an armed terrorist group seeking a pathogen. On the other hand, a facility possessing a virulent form of *Bacillus anthracis* (the causative agent of anthrax) could face a serious threat from armed extremist groups that have no interest in research animals. Both threats exist, but the risk of occurrence and consequence will drive the determination of which security measures provide the best realistic protection. An understanding of the potential threats, capabilities, and motivations of those who would carry them out provides another building block in the physical security development process.

The Vulnerability Assessment

The final building block, often overlooked, is a vulnerability assessment (VA). The VA looks at the security electronic systems, physical features (i.e., barriers), building design, operational controls, and protective forces to develop a comprehensive analysis of security effectiveness. Like the threat assessment, a VA is best accomplished by using subject matter experts (SME) in fields such as security, science, maintenance, building designers, and management. Other key assets are persons responsible for the security and safety of the facility. For a university this could be the campus police or, in other situations, a local law-enforcement representative. These individuals know the available resources, training needs, and response capabilities of the facility. The same applies to medical services and fire department personnel. Through this multidisciplinary approach, conflicts between security and other stakeholders are identified and resolved so science and maintenance operations are not burdened with unnecessary security requirements yet security needs are ensured. During new construction or renovation, the VA should be conducted at both the design stage and after completion. Postconstruction analysis ensures that the final design and construction remains compatible with the VA. Figure 4.1 demonstrates the relationships between risk assessments, threat assessments, and vulnerability assessments in the contexts of both biosafety and biosecurity.

Risks presented to laboratory workers and the public are presented *from* the biological agents/VBM. Threats are presented *to* the biological agent/VBM. Vulnerability assessments can be applied to address vulnerabilities in both risk and threat assessments. Note that the risk assessment is a fundamental component of biosafety and the threat assessment is a fundamental component of biosecurity. There is overlap between biosafety and biosecurity, as described in Chapter 2.

The end goal of these assessments is to develop a security program with a physical component that will detect attempts at unauthorized access to BSAT, facilitate assessment of the threat, create a delay, and provide for a response

FIGURE 4.1 Assessment relationships

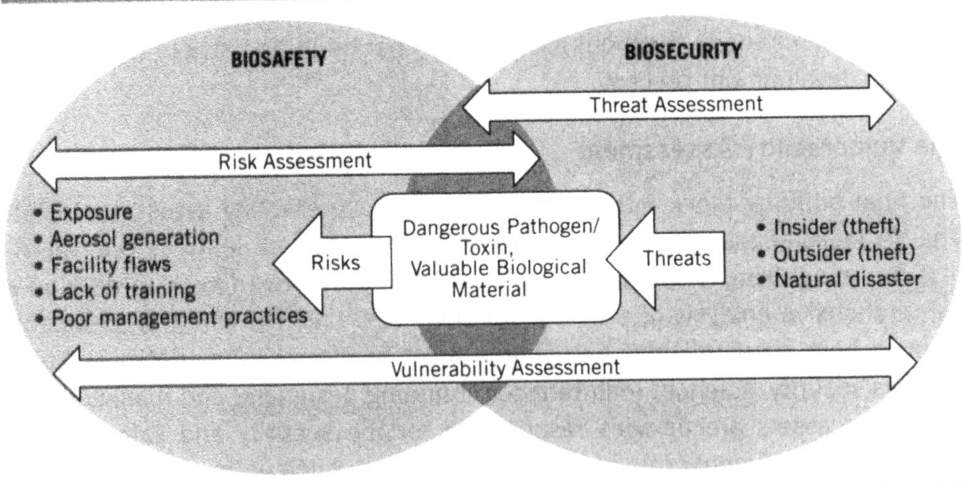

FIGURE 4.2 Layered security

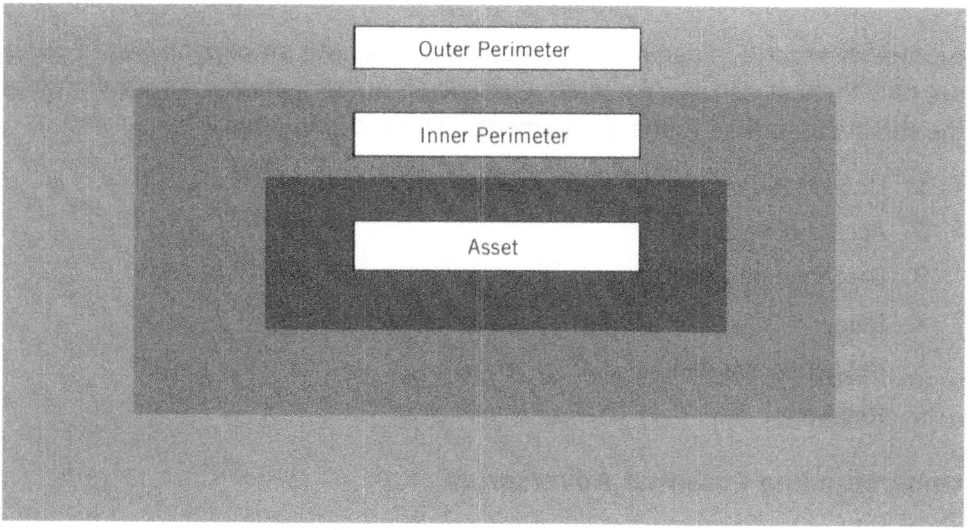

capable of neutralizing the threat. At the same time, care must be exercised to avoid implementing requirements that are unreasonable in their application and that create an unnecessary impediment to the needs of science. This type of security is achieved through a "layered" or "graded" approach in which security measures are applied in increasing increments the closer one gets to the select agent (Figure 4.2). This "layering" is seen in access control and in the use of physical barriers.

Each facility will have its own unique characteristics that drive the levels and technology or design features used in the security program. For example, a veterinary diagnostic laboratory in the rural southwest may be a standalone building resting on a 10-acre tract of land. A research laboratory, on the other hand, may be located on the third floor of a university campus building, sharing floor space with classrooms and offices. The concept, however, remains the same as illustrated in Figure 4.2.

As with access control layered security (a layered configuration) is the optimal approach, beginning with the perimeter of a facility and working inward to the building "skin" or walls, then to the area, often referred to as a biological restricted area, where pathogens or other assets are physically located.

BREAKING DOWN THE PROGRAM: THE SIX SECURITY CAPABILITIES

A physical security program must effectively integrate six capabilities in order not only to meet certain regulatory guidance requirements but also to mitigate the threats identified in the assessments. These capabilities are:

1. Understanding potential adversaries
2. Access control
3. Detection and assessment
4. Delay
5. Response and denial
6. Recovery

Understanding Potential Adversaries

Understanding which groups or individuals pose the greatest risk to the institution begins with the threat assessment (see Chapter 6). The assessment has defined groups of potential "adversaries" both from an insider and from an outsider perspective.

The insider is commonly thought to be motivated by financial gain, ideology, or as a victim of coercion, while the outsider is thought to be motivated only by financial gain or ideology. The outsider may act alone or may solicit active or passive assistance from an insider. A passive insider could merely provide information, such as how many security officers are on duty or where a specific lab is located, whereas an active insider may bypass security systems or create a diversion.

The assessment also identifies how motives are accomplished. These can include:

- Information theft
- Theft of material
- Disruption of operations
- Release of material

The Insider Threat

Often the most dangerous threat comes from within. This is not to say that those who work with biological materials are automatically suspect. These personnel know and are familiar with the virulence and pathogenicity of specific pathogens and have access to security protocols of the facility. This concern over personnel responsibility/accountability was reinforced in 2001 when 5 people were killed and 17 injured by anthrax spores mailed to multiple recipients. While the actual source of the spores continues to be debated, there is a strong belief that the strain originated in a lab in the U.S and was disseminated by someone with authorized access (see Chapter 3). As a result, there has been a renewed focus on background checks for those who work with select agents and on potential enhancements to current requirements. Possibilities include a more in-depth background check supported by recurring checks—similar to those required for security clearances. The rationale for periodic reassessments is that, like pathogens, people who work with them are ever-changing, meeting new stresses and life experiences, which can have undesirable effects on reliability.

The Outsider Threat

The outsider can wreak havoc with research, create lack of confidence in facility safety, and, in the extreme, pose a danger to life and health. The outside threat may be simply motivated by a quest for financial gain or a dedicated "bioterrorist," with the sole intent of releasing a virulent pathogen on the public as an act of aggression. While the first response may be to think that both persons are acting criminally, there is a distinction between the two. The criminal intent on financial gain is looking not to destroy but rather to obtain something of value. An outsider acting on ideology may or may not be concerned with destruction or loss of life, his own or others. Granted, these categories may cross lines, but the ideologist could be considered the greater hazard, depending upon the level of commitment, expertise, and backing.

Potential scenarios might include a criminal with the intent to steal a pathogen for sale or for ransom. This type of person is not likely to be a dedicated bioterrorist nor is he likely to have the training needed to mount a full-scale attack. He would likely choose a covert approach and remain undetected in order to avoid physical confrontation.

In another scenario animal-rights activists want to draw attention to what they perceive as mistreatment of research animals. This group poses multiple risks to the public. At a minimum, the potential loss of research data and public confidence would have a detrimental effect on research efforts. A particular activist could break into a lab and in a misguided sense of compassion release the animals, not realizing they may carry pathogens and pose a risk to public health. As with the previous example, this person would prefer stealth to discovery until after the fact, at which time she will claim recognition.

In a thrd scenario a dedicated, well-trained, and well-equipped bioterrorist has no qualms about loss of life, his own or others. This person could be expected to have conducted extensive research in order to understand where his target is located and how it is protected, and he is prepared to attack in force. The goal may be to obtain pathogens or to release pathogens, thus causing civil panic, loss of life, and disruption of business.

Access Control

Access control is quite simply defined as knowing who or what can enter specific areas of your facility and who or what cannot and how that access is regulated. Federal regulations require that a person be approved for access to BSAT, but there are other factors to be considered when determining who is to have access and to where. A well-protected facility will have a procedure in place that identifies who is allowed to be on the premises and under what conditions. For example, staff members who passed mandated background checks may be authorized to have access to biological material and to any rooms where material or related data are stored. Those staff members whose duties do not require access to material or data may be restricted as to what areas of a facility they can enter unless they are escorted by a qualified person. Likewise, visitors must be categorized and provisions developed to regulate their presence and actions.

Controlling the entry and egress of authorized personnel is accomplished through a variety of methods. These can include authorization lists, handwritten entry/exit logs, electronic keys with embedded data that allow predetermined access, and the issuance of hard keys. In some applications the use of biometric identifiers such as iris scanners or hand readers can provide a higher level of security that may prove more user-friendly than manipulating key cards or remembering a PIN code. An ever-tightening control of access illustrates the concept of "layering" from an access control perspective. Unfortunately, even the most stringent access control measures may not identify the insider threat.

The uninvited are those who have chosen to come onto your premises or into areas where they are not authorized without your permission and as such must be assumed to have unfriendly intentions, as identified by the threat assessment. It is these uninvited and unwelcome persons who pose a threat of theft and sabotage through overt or covert acts. When confronted with an outside threat, conventional access control measures become less relevant, and mitigation relies on the next capability of detection and assessment.

People are not the only concern of access control. Deliveries, materials, and vehicles are equally important. The topical areas that should be addressed in an access control procedure are discussed later in this chapter.

Detection and Assessment

Detection is the process of recognizing an unauthorized or attempted unauthorized act. As a matter of practicality it is best to always assume detection includes assessment. Consider a standalone building, unoccupied at night, with a door alarm. The alarm will sound (detect) if someone breaches the door. But if no one is assessing the cause of the alarm, it may not have been detected.

Detection/assessment can be achieved by a number of means including human observation and a multitude of electronic sensors. Due to the many types and applications, the appropriate electronic systems require a skilled specialist to identify the best equipment and placement to meet a specific need. In general, however, these include but are not limited to microwave and infrared sensors, used to detect movement and penetration around doors and windows. Closed-circuit television (CCTV) can be integrated into the detection capability so that, upon receipt of an alarm at a monitoring station, a camera will automatically provide assessment capability.

Delay: Keeping Out the Uninvited Until Help Arrives

Delay features are used to deter and, when needed, to delay an adversary access to an area pending arrival of a response capable of neutralizing the threat. Physical barriers are the most commonly used delay. The type of barrier and its construction are predicated on a number of factors, one of which is the time needed for appropriately equipped responders to arrive in sufficient numbers. As part of the VA process, the amount of delay needed can be calculated to determine the level of barrier needed. Consider a facility is surrounded by an eight-foot chain-link fence topped with concertina wire and monitored by intrusion detection and CCTV assessment. Through analysis it is determined that an adversary can reach a desired target location in five minutes from the point of detection at the fence line through an exterior building door, ascending two flights of stairs, and breaching a metal door. That analysis also reveals that the on-site response force requires six minutes to arrive at that same location. Using this analysis, it can be determined that increasing the delay time by a minute or more will provide the responders the needed time to be in a position to deny the adversary access. This delay may be achieved by installing a heavier door at the target location, a second perimeter fence, or additional barriers on the adversary pathway. Conversely, an analysis may determine that it is more cost-effective to relocate the responders so their transit time to the target area is decreased. In the "real world" application, delay tactics include a combination of increased delays and shortened response times.

The most common feature around a facility is fencing or walls, serving both as a demarcation of boundaries and a means of directing personnel and vehicles to specific entry points. The design is driven by the findings of the threat and risk assessments, physical characteristics, and location. While it is not practical to build a fence around a third-floor research lab, the threat and risk assessment at a standalone BSL-4 lab may justify the use of a high-security fence with controlled gate access. Fencing and walls can be supplemented by architectural features, such as fixed obstacles like hydraulic-powered barriers that can rise from a buried position in tenths of a second to create a vehicle barrier. Aesthetically, such obstacles can even be disguised as planters or statues.

The outer walls or "skin" of the building represents the next protective layer and can include interior walls. Exterior walls are a deterrent by virtue of their material—i.e., brick, stone, concrete. Therefore, the overlying concern is ensuring that access points to the building are protected in a manner sufficient to deter and delay a breaching attempt. While doors and windows are the most obvious targets, a thorough VA will have identified alternative pathways such as those afforded by the ventilation, plumbing, and electrical systems. Each of these paths must be assessed to determine their viability and the level of protection required to mitigate the ability of an adversary to utilize them as a point of undetected ingress.

The biological restricted area, or the immediate area where assets (data or pathogens) are located, is the final layer of protection. Here the most robust systems, barriers, and access controls, are needed to ensure that only those with "need for access" can gain entry. If unauthorized personnel somehow access a biological restricted area, then an alarm or other response should occur.

Response and Denial

In spite of all the access controls and delay features in a security design, there is still the chance that someone may actually breach a layer of security, thus posing a threat. Therefore, an effective response either interrupts an unauthorized act before it is performed or prevents the perpetrator's escape. Once again, the preliminary assessments guide the response capability. In some instances, it may be sufficient to rely on local law enforcement to respond. In more extreme situations, it may be necessary to employ on-site security to ensure a timely and successful response.

Recovery

The recovery mission from a physical standpoint may fall to a law enforcement agency rather than the facility, but the facility must be able to effectively communicate to first responders, whether law enforcement, medical, fire department or public health workers, the specific agent which has been compromised, any precautions required and as much information as can be provided on the circumstances surrounding the loss or compromise. Additionally, there must be a documented and proven procedure in place to drive the response.

ARCHITECTURE, ENGINEERING, AND WORK PRACTICE CONSIDERATIONS

The practical application of physical security in a biosecurity setting requires collaboration between the security SMEs, building design engineers, architects, and end users. This collaboration is essential to effectively integrate security into building design features and existing work practices so each discipline is mutually supported. Just as security SMEs expect other team members to assist in finding solutions to security concerns, the security SMEs must be cognizant of and responsive to the needs of the end users and the mission. A security system designed in a vacuum risks noncompliance with critical security measures and unnecessary impediment of the facility mission.

Collaborative and Complimentary Design

As the construction design is developed, architectural designs and construction material specifications should be reviewed by security SMEs to ensure that there are no conflicts with security best practices or analysis findings. For example, during design development for a new lab it was discovered that, due to fire protection codes, security doors for certain rooms could not be installed in a manner that would provide optimum protection. The concern was raised with the building designers. As a solution, the rooms were re-designed with vestibules, allowing the doors to be installed as desired. In another instance, a review by security analysts revealed the planned landscaping design included trees adjacent to a perimeter fence. The trees provided not only concealment but also a means to cross the fence without detection. The client was adamant about keeping the trees. By identifying the conflict early in the design, additional security measures were incorporated to mitigate the vulnerability. Parallel design and review also can reduce costs by suggesting shared cost areas such as conduit, trenching, lighting, and other infrastructure features.

Design engineers and facility staff—for example, biocontainment specialists and scientists—are also allies in designing the security system. Through consultation, the team can work together to resolve security concerns and develop designs that complement security. In one biological containment facility, there was a concern about penetration into a room due to the presence of an autoclave large enough for a person to pass through. Due to space restrictions, there were no design alternatives that would facilitate the installation of electronic security detection devices. After consultation with the designers and end users, it was discovered that, due to the interlocking features of the autoclave and work practices, the pathway was not viable and the issue was resolved.

As mentioned previously, personnel responsible for emergency response should be included in design discussions. A facility may have a security department and an existing alarm monitoring system that can integrate the system developed for the biosecurity program. The same holds true for medical and fire responders. They must be aware of what changes are coming and have the opportunity to identify conflicts with current practices. This collaborative approach does not stop with the conceptual design stage but is a dynamic process requiring ongoing interface throughout not only building design but also security system design. In fact, this collaboration should take place anytime there is a change in configuration or remodeling. Electrical loads, for example, need to be shared with designers as well as specifications for doors, walls, and windows to ensure materials meet the requirements of the security analysis. As a design progresses through the review stages, all team members should review changes that have occurred to ensure that there are no negative impacts to the security system, the budget, or the ultimate facility function.

Access Control

A component of access control is the knowledge of who has access and how they are controlled. The building design should incorporate a layout that places common areas such as meeting rooms, break rooms, and administrative offices in an outer layer accessible from the exterior of the building. Further access to corridors or floors that house or lead to valuable assets such as data or pathogens should be limited to persons deemed to have a need to work in those areas or under procedurally identified requirements for escorted access. The third layer, where the asset is actually located, should be under stringent access controls that may include but are not limited to two-person access, dual verification at card readers, or differently keyed dual padlocks.

In multifloor buildings, a design may allow for all administrative functions to be conducted on the first floor, with laboratories and other sensitive areas on upper floors. For example, in a three-story building, one might find a design that would accommodate common areas on the first floor, with access to the second floor limited to those authorized to work in a BSL-2 environment. The third floor correspondingly would be limited to BSL-3 activities and those authorized to work in those conditions. This is not a "hard and fast" recommendation but rather a simplistic view of layering.

Another design feature worth considering is consolidation. This does not mean that everything must be in one room or building but is rather a working concept to minimize the locations where assets are held. This concept applies to where pathogens and high-value targets are located. Consolidation can:

- Reduce the number of areas requiring emergency response
- Decrease response times by decreasing distance
- Yield cost savings by reducing need for security equipment

While not always feasible, the principle of layered security is more readily implemented by funneling personnel through ever-tightening restrictions to a tightly concentrated area.

Barriers

A "layered" barrier design supports access control layering by allowing routine, non-sensitive operations while deterring and/or delaying unauthorized personnel. As noted above, the layers start at the property line/outer boundary and progress inward until the asset being protected is reached. When designing barriers, especially doors, windows, and walls, it is critical to ensure that the delay afforded by one is commensurate with all. For example, it serves no purpose to install a steel door into a reference museum if the walls are made of 3/8" sheetrock that can be easily broken. Walls can be reinforced in a variety of ways, including the installation of light-gauge wire mesh inside the wall or by using penetration-resistant material. In some cases the walls may even be of concrete blocks. Likewise, windows should be impact-resistant, with laminate or reinforcing wire being a common feature.

While walls, doors, and windows may seem like obvious points of entry, to the determined outsider there are other less obvious routes to exploit. Among these are heating and ventilation (HVAC) ducts, utility chases (trenches) for piping and electrical conduit, exterior ladders, and rooftop doors. These should be identified in the vulnerability assessment and a mitigation strategy of barriers and detection/assessment developed. Exterior ladders can be mitigated by not designing them to reach to the ground but instead accessing them from an upper level or, if at ground level, securing by a substantial locked, alarmed, and monitored gate. Utility chases and HVAC ducts can be secured by grates and pipe baffles and monitored by a variety of detection systems.

Detection and Assessment

Once the layers are established, they must be monitored in order to be effective. An eight-foot-tall fence topped with concertina wire is virtually ineffective if not monitored. At the fence line, monitoring is accomplished through an array of technology including microwave sensors, infrared detection, ground sensors, and fence vibration sensors. Doorways are easily monitored by balanced magnetic switches (BMS), which can be accessed during normal business hours for routine access points, and windows can be monitored via glass break sensors. These should be installed in rooms containing assets, in the corridors leading to them, and in ductwork and utility chases; sensors such as motion detectors can be a solution.

One may see assessment capability treated as a standalone element, but as mentioned earlier including it with detection ensures that it is a complementary feature. If one receives an alarm and no one investigates the cause or looks at the point of alarm for the cause, the alarm itself is ineffective. Assessment can be as simple as sending a person to investigate a sound, or it may incorporate high-tech cameras with pan/tilt/zoom (PTZ) features, video motion detection that both senses movement and calls up a camera image, or fixed cameras monitoring a hallway.

As with all technology, there are a multitude of products on the market, and each of them has a use that benefits a specific design. This chapter is not intended to discuss the individual merits of or recommend a specific approach. This is where the work of SMEs in security VA and design are essential to ensure that the physical security design is effective both from a security standpoint and for the user.

OPERATIONS AND SUSTAINMENT

As one draws near to completion of the security design, the realization comes that, beyond building it, installing it, and starting it up, there will be a need to both operate the system in accordance with its intended design and to sustain it. Both operation and system sustainment come with challenges and needs.

Challenges

One major challenge faced by security professionals is the reluctance of end users to comply with identified security measures. This does not imply that the users deliberately circumvent security with malevolent intent but rather that they do not understand the reasoning or do not like the restrictions incurred. There have been cases where security doors were propped open simply because the occupants of one room wanted to converse with those in another room, yet this was not only in violation of security protocol but also lab protocol. This is one example demonstrating the importance of early end-user input into security plans.

Another challenge is associated with interfacing security/law-enforcement and life-safety personnel in the emergency response. In some instances, there may be one individual responsible for biosecurity and another responsible for response to nonbiosecurity incidents. There is justification for this approach since biosecurity and biosafety are focused differently, but both disciplines will find their programs enhanced by cooperation (see Chapter 2).

Designing for Requirement-Based, Sustainable Solutions

One effective method to gain cooperation and buy-in from laboratory end users is to keep the program requirements basic. This means avoiding the temptation to use the "latest and greatest" technology, which may not be practical. It could be very tempting to go out and buy the latest retinal scanners on the market, but if the job requires low-tech goggles, then security has created its own noncompliance issue. Conversely, if a keypad is installed where a worker is commonly going to be wearing gloves that restrict agility, retinal scanners may be the better value from a compliance perspective.

The program must also be sustainable from the perspective of equipment and of the persons tasked with implementing the program. If the end user has limited computer skills, the margin of error is greatly increased. For example, one plan involved the use of a computerized access control system at the main entry point to log visitors in and out. It was discovered prior to purchase, however, that the protection force would not likely receive the training necessary to operate the system. Follow-up assessments indicated that end users fared just as well using handwritten logs, and they displayed a sense of detail and pride when the logs were audited for accuracy and compliance.

Parts availability for security equipment is also a sustainment issue. Electronics used should be locally available to the extent possible, and, if not, an inventory control system should be established for items requiring delivery from external sources. This can be extremely critical in some underdeveloped or emerging countries.

Security Plans and Procedures

Finally, there must be a mechanism to ensure that the program is implemented consistently and in compliance with regulatory guidance. Management, administrators, and employees will come and go, but the facility and the security program will always be there. It is essential to develop a comprehensive security plan that defines the overall philosophy and administrative responsibilities for training, implementation, sustainment, and compliance. This plan must be supported by procedures that describe the actions necessary to implement the various parts and who is responsible for those actions.

Access Control

The access control procedure describes who has authority to determine who gets unescorted access versus escorted access and how that authorization is communicated down the line of authority. The procedure also contains direction on which steps are required for personnel, vehicles, deliveries, and personal items to enter and exit the site. This includes but is not limited to search requirements, dropoff points, and access to data if not addressed in an tnformation security procedure.

Key Control

The key control procedure establishes the control criteria for the issue of hard keys, key cards, and personal identification numbers (PINs). It will identify who can authorize receipt of these items, who can issue them, the requirements for issue and return, and a process for key accountability and audits to identify loss or compromise.

Response

In order to ensure a consistent response to incidents, a response procedure will identify who has overall command of a response, who has on-the-ground command, and what are the responsibilities of responders. The procedure includes all the potential risks and threats identified in the assessments and contains direction on the actions to take for a specific incident. The incorporation of an off-site response must also be addressed.

Compensatory Measures

If not included in the response procedure, a separate document should be developed to address compensatory measures in the event of a degradation of security such as loss of lighting, loss of alarm capability, or loss of keys.

System Testing

Provisions must be established for testing the security system's electronic components, operability on a daily basis, and full system compliance on a less frequent schedule. This testing must include the electronic access controls, the camera system, and the alarms and the interface among them.

Quality Assurance and Compliance Assessment

The quality assurance (QA) process provides a vehicle to ensure that the program is operating as intended through scheduled and unscheduled assessments. In addition to testing the system, personnel are also assessed through work performance observations, training exercises, and drills.

Training

Prior to assignment to any role, each staff member must be made aware of his role and responsibility for the safety and security of employees, the public, and the facility. The training procedure will define the training hierarchy, the requirements for initial training and updates, and, when needed, remedial training. A tracking process and training documentation methodology are also key components. This training should include emergency responders. Periodic exercises should be scheduled with emergency responders using realistic scenarios so that all parties become acquainted with one another, familiar with one another's roles and expectations, and familiar with physical layouts.

CONCLUSION

The success of an effective biosecurity program relies on an integrated, sustainable, and compliant program: integrated, in that it considers all users and incorporates complimentary design, operations, and security features; sustainable, in that it is practical and requirements-based; and, perhaps most important, compliant, in that it not only meets regulatory requirements by letter but is also supported by the actions of those regulated.

Incorporating physical security into biosecurity is not a mystery nor is it a process that can be developed in a vacuum. Ideas and solutions come from both expected and unexpected resources. Sometimes the building engineer with the least familiarity with security will provide the solution to a concern. Security professionals will often find themselves at odds over what later are seen as trivial matters such as terminology as well as over the best solution. At the end, however, the result is a program that accomplishes the key capabilities of: understanding, controlling, detecting and assessing, delaying, responding and recovering.

References

1. Interagency Operation Security (OPSEC) Support Staff (IOSS). National OPSEC Program. Glossary of terms. Accessed at: http://www.au.af.mil/au/awc/awcgate/documents/opsec_definitions.htm.
2. Alibek, Ken, and Handelman, Steven. Biohazard: The Chilling True Story of the Largest Covert Biological Weapons Program in the World—Told from Inside by the Man Who Ran It. Random House. New York (1999).
3. Intellectual Property Rights Center. Joint Strategic Plan on Intellectual Property Enforcement (2011). Accessed at: http://www.iprcenter.gov/reports/ipr-center-reports/2011-joint-strategic-plan-on-intellectual-property-enforcement/view.
4. Department of Health and Human Services. Centers for Disease Control. CDC Select Agents Regulations. Federal Register 42 CFR Part 73 (2008).
5. U.S. Department of Agriculture. Animal and Plant Health Inspection Service. APHIS Select Agents Regulations. Federal Register. 7 CFR Part 331 and 9 CFR Part 121 (2008).
6. U.S. Department of Health and Human Services. Centers for Disease Control and Prevention. Biosafety in microbiological and biomedical laboratories, 5th ed. HHS Publication No. (CDC)21-1112 (2009).

CHAPTER 5

Operational Elements of Biosecurity

J. Craig Reed and Debra C. Sharpe

INTRODUCTION

Biological organisms and their products possess unique properties that necessitate a different application of security beyond mere physical security (Chapter 4). Under the proper conditions, biological organisms will grow and multiply. From a security perspective this is problematic because, theoretically, only a single organism may be required to produce infinitely more organisms. This is not inconsequential, as some organisms are pathogens known to be highly dangerous to human and/or animal health. Toxins also necessitate biosecurity. Toxins are biologically active molecules that interrupt normal cellular processes and are produced by some organisms. Toxins elicit adverse consequences on the cells, tissues, and organisms to which they are exposed. Furthermore, issues of intellectual property and real or perceived financial value are of concern because some organisms found in research and commercial environments possess unique attributes as a consequence of genetic engineering and recombinant DNA techniques. These various properties of biological materials necessitate a unique approach to security that begins with the understanding that biological security, physical security, and biological safety are intimately connected and must be mutually supportive for maximum effectiveness.[1] Furthermore, professional staff with management responsibility for these disciplines must communicate frequently and openly with each other not only to ensure that each of these areas is fully integrated with the others but also to ensure that all parties have visibility and timely awareness of issues within each area.

Operational biosecurity is the implementation of control measures necessary to safeguard biological material of commercial value (including genetically modified microorganisms, unique research animals, recombinant DNA constructs and/or protein products derived from them), high-consequence pathogens, biological toxins, and/or intellectual property associated with any of these materials. Threats to these biological materials fall into two broad categories, the outsider threat and the insider threat, as explained in detail in Chapter 6. These threats must be documented in a threat assessment and thoroughly analyzed and prioritized based on the likelihood and consequences of loss, theft, or misuse of the biological materials. The threat assessment and the risk-based prioritization of the threats to the organization must be documented in a formal biosecurity risk assessment. This risk assessment drives the creation and implementation of a variety of administrative controls, including a biosecurity program plan, which in turn provides the foundation for a biosecurity program to manage and mitigate the threats identified in the risk assessment.

THE BIOSECURITY PROCESS

The term "biosecurity" has several definitions. In animal settings, where livestock or colonies of laboratory animals are managed, it relates to the protection of the animals from the spread of infection by zoonotic disease agents. The use of the term in this chapter will refer to the U.S. Centers for Disease Control and Prevention (CDC) definition in the 5th edition of Biosafety in Microbiological and Biomedical Laboratories (BMBL). In Section VI, biosecurity refers to the protection of microbial agents from loss, theft, diversion, or intentional misuse. The mechanics of how organizations ensure the protection of these agents and toxins is the biosecurity process. This process includes the risk assessment; development of a management plan to mitigate the risks; evaluation of personnel to ensure their suitability to work with these materials; control and accountability for the manipulation and storage of high-consequence pathogens and toxins; physical facility access and control elements; secure material storage; use, and transfer documentation; biological safety and biological security training; emergency response training; and ongoing program management.

Risk Assessment

The first step toward implementing the concept and practice of biosecurity is to conduct a biosecurity risk assessment. A biosecurity risk assessment is quite different from a biosafety risk assessment. A biosafety risk assessment is fundamentally focused on worker safety and is principally concerned with the risks presented to human health by the microorganisms in the laboratory. The biosecurity risk assessment, on the other hand, is focused on the security of the microorganisms and is concerned with the risks presented by those individuals who may wish to remove or release them from the facility. The biosecurity risk assessment is the product of a threat assessment (i.e., the identification of threats) followed by an estimate of the risk associated with each threat (i.e., the likelihood of occurrence of any given threat event and the severity of impact or consequences associated with that event). It is this estimate of risk that prioritizes threats. The biosecurity risk assessment is therefore composed of four steps: (1) identification of critical assets within the organization, (2) identification of the full range of credible threats to those assets, (3) a subjective determination of risk associated with each threat, and (4) prioritization of the identified threats based on the associated risk. It is important to note that the biosecurity risk assessment must be reperformed periodically (1) to ensure that new and emerging threats are identified in a timely manner and fully considered through the threat assessment process and (2) to determine whether the risk associated with each threat has changed as a result of new policies, procedures, or technologies.

The risk assessment begins with the identification of those assets that, if stolen or damaged, would cause harm to others or the environment or would compromise the operation of the organization. These assets may include cultures of select agents[2] or hazardous microorganisms; animals or plants infected with these pathogens; animals or microorganisms that possess unique genetically engineered attributes; intellectual property (such as proprietary process information); floor plans revealing the location of sensitive materials, access control points, or the location of critical elements of facility infrastructure; and sensitive administrative control documents such as emergency response procedures or incident response plans.

Once the assets of the organization are identified, all possible threats to each asset must be identified and prioritized through a process called threat assessment. Threats generally fall into two broad categories, depending upon their origin: insider threats and outsider threats. The insider threat is a threat originating within the organization by someone who possesses authorized access and has both the intent and the capability to either compromise or damage facility infrastructure or to steal and cause harm with the biological materials. The outsider threat encompasses any threat originating outside the organization. Outsider threats include individuals without authorized access but with both intent and capability to compromise or damage facility infrastructure or steal biological materials and use them to cause harm. Outsider threats also include meteorological events (such as tornadoes, hurricanes, uncontrolled fires, and flooding), geological events (such as earthquakes, volcanoes, and tsunamis), and riots and mass gatherings.

Through the process of threat assessment, the insider and outsider threats to the organization must be thoroughly identified, rigorously analyzed, and cautiously prioritized based on the risk each presents. Risk is calculated as the product of the likelihood of threat occurrence and the adverse impact or consequence to the organization if the threat occurred. For biosecurity purposes, adverse impact or consequence to the organization typically involves the loss, theft, or misuse of the assets or an inability of the organization to fully conduct its normal operations following a threat event.

As discussed in the next chapter, the identification, analysis, and prioritization of the insider threat benefits from the involvement of a qualified threat assessment professional as well as properly trained physical security personnel and/or mental health professionals (Chapter 6). These individuals, along with senior management and security representatives, must possess an understanding of the targeted violence process. In order for the organization to periodically reperform its threat assessment, a combination of background and suitability investigations is needed so behavioral baselines can be established for new hires. This must be coupled with a threat assessment investigation for existing employees that display threatening behavior toward other employees or the organization.

External threats can best be identified and prioritized with the support of local and state security organizations. In the U.S. this would include the Department of Alcohol, Tobacco, and Firearms; Federal Bureau of Investigation; Department of Homeland Security; and Department of Commerce. These organizations are instrumental in the prioritization of identified threats. The outcome of a properly conduced biosecurity risk assessment is the identification of all insider and outsider threats to the organization and an understanding of their associated risks. In the written assessment, the threats are prioritized (from most significant to least significant) and accompanied by a brief description of the associated impact to the organization. The risk assessment must be shared with senior management, safety, and security professionals within the organization, and select other individuals who are involved in the development and implementation of the biosecurity program plan. Awareness of the risk assessment by management and professional staff permits them to identify risks that are unacceptable and properly prioritize resources as they mutually define the approach to the management or mitigation of the remaining risks. These management and mitigation approaches must also be defined in writing within the biosecurity program plan. It is senior management's role to distinguish between threats with unacceptable and acceptable levels of risk.

Administrative Controls

The hierarchy of administrative controls begins with federal regulations, guidelines, and internationally accepted best practices. Within any organization that implements or adheres to these regulations, guidelines, and practices, these documents drive the creation and issuance of multiple policy statements. These policy statements emanate from the office of the organization's director. Implementation of policy directives necessitates the creation of one or more individual programs, each of which is guided by a formal written program plan. Successful program execution is accomplished through identifying all of the individual practices and procedures that must be performed by staff members. The process of performing these activities is standardized and documented in a standard operating procedure (SOP). Standard operating procedures ensure performance consistency and uniformity among different individuals. Staff compliance to SOPs is achieved through the creation and delivery of training content specifically focused on the critical tasks and performance elements of each SOP. To ensure maximum effectiveness, training events must employ adult learning techniques, appeal to a variety of learning styles (i.e., visual, auditory, experiential), and identify specific training outcomes or competencies (i.e., knowledge, skills, and abilities) that are both observable and measurable. Critically, training must reoccur frequently to ensure sustained compliance.

Administrative controls (or work practice controls) are changes in work procedures such as written safety policies, rules, supervision, schedules, and training with the goal of reducing the duration, frequency, and severity of exposure to hazardous "situations."[3] Consequently, at the level of the organization, administrative controls represent the practical application of federal regulations, guidelines, and internationally accepted best practices within a given organization. Truly effective administrative controls are internally developed and specifically address unique activities, process flow, and work environment of the organization. The development of these documents is an organic process and unique from organization to organization.

Administrative controls form the foundation of effective biosecurity. The minimum controls include a biosecurity policy statement, a biosecurity program plan, and a variety of biosecurity SOPs. All of these controls emanate from the biosecurity risk assessment and must be modified as the assessment of risk changes. The creation of this hierarchy of controls by members of the organization reinforces the value and importance of open communication and the sharing of information across disciplines, team-based identification of risks and approaches to management or mitigation, and the development and delivery of biosecurity training to staff members. Annual review of all these administrative controls is essential. This procedure ensures that policies and procedures of the organization remain consistent with the activities of the organization. It also ensures that these controls remain properly aligned with the findings of the most recent biosecurity risk assessment. The annual review also ensures that recommended procedures are updated to reflect changes in technology or are eliminated due to changes in the organization's activities.

Biosecurity Policy Statement

The biosecurity policy statement is a clear and succinct statement from the senior leadership of an organization that communicates to staff members the importance of biological security for the protection of a variety of valuable and/or hazardous biological material as well as intellectual property and electronic information. Specifically, the biological security policy statement addresses the organization's: valuable biological materials, including select agents and toxins as well as proprietary materials and reagents; proprietary process information, research protocols and intellectual property; facility floor plans that reveal the general or specific location of areas where biological materials are manipulated or stored, where animals are housed, or the location of security devices; security measures that do or don't exist within the facility; incident response plans; and the existence or location of critical elements of infrastructure. The policy statement must clearly articulate support for all existing regulations (federal, state, and local) as well as guidelines and best practices pertaining to biological security. For organizations within the United States, depending upon the nature of the work performed, this likely includes reference to the U.S. Patriot Act of 2001, the Public Health Security and Bioterrorism Preparedness Response Act of 2002, U.S. Department of Transportation regulation 49 CFR Parts 171-178, Public Health Service regulation 42 CFR Part 72 and 73, 7 CFR Part 331, 9 CFR Part 121, the latest edition of the International Air Transport Association Dangerous Goods Regulations, the latest edition of the National Institutes of Health and Centers for Disease Control and Prevention joint publication entitled Biosafety in Microbiological and Biomedical Laboratories, the CEN Workshop Agreement (CWA) 15793:2011 and CWA 16393:2012, and, possibly, the Biological Weapons Convention Anti-Terrorism Act of 1989, the Animal Health Protection Act, the Plant Protection Act, and/or the International

Health Regulations.[4,5,6] Organizations outside the United States should reference similar federal and state regulations as well as the latest edition of the World Health Organization's Laboratory Biosafety Manual.

To ensure that staff and visitors understand the importance of biosecurity, the biosecurity policy statement must be signed by the director of the organization and placed in a public location near the main entrance of the facility in close proximity to the organization's safety policy statement.

Policies should limit access to high-consequence pathogens and toxins to authorized individuals. Organizations should consider developing polices to identify and report suspicious behavior. Policies of this nature should always involve the human resources department to ensure that they are written and applied in such a manner as to avoid discrimination, that there are no violations of federal or state laws, and that they are implemented fairly across the organization.

Biosecurity Program Plan

The purpose of the biosecurity program plan is twofold: to identify all insider and outsider threats to facility infrastructure, biological materials, processes, and intellectual property within the facility and to identify a graded approach to the mitigation of these threats. For maximum effectiveness, the biosecurity program plan must be developed with direct input from and involvement of a variety of different stakeholders and subject matter professionals including:

- Senior management
- Human resources officials
- Scientific staff
- Veterinary staff
- Responsible official
- Information technology staff
- Housekeeping staff
- Physical security expert
- Biosecurity expert
- Biological safety expert
- Facility operations and engineering staff

Each of these parties brings unique and valuable contributions to the biosecurity program plan due to their varied perspectives and awareness of issues directly related to the security of biological materials and intellectual property. Scientific staff members are well aware of which organisms possess significant value from an intellectual property standpoint and which pose the greatest risk to people, animals, and the environment. Veterinary staff members know where infected animals are housed and how infected carcasses are moved, stored, and treated. The responsible official (RO) knows the locations where BSATs are handled, manipulated, and stored. Senior management knows which individual researchers are performing sensitive work with high-consequence pathogens and toxins. Housekeeping staff members may handle potentially infectious waste materials before and after decontamination or store them until a contractor removes them from the property. Facilities operations and engineering staff operate and maintain the critical facility infrastructure responsible for maintaining containment, filtration of air, and processing of biological waste in solid and liquid waste streams. The combined involvement of these individuals in the process of biosecurity program plan development will ensure that the final plan is simultaneously comprehensive, broadly applicable, and widely endorsed across the organization. This approach decreases the likelihood that important elements of the program plan are inadvertently omitted or insufficiently addressed, and it ensures the consistent execution of the program plan across the organization.

The biosecurity program plan should also identify all extra-entity agreements with local law enforcement and emergency response teams such as fire and hazardous materials responders. When outside teams are used to respond to emergencies, the entity must identify how it will ensure the security of high-consequence pathogens and toxins from responders.

Biosecurity Standard Operating Procedures

Standard operating procedures ensure consistency and uniformity in the performance of the same task by different individuals. Biosecurity SOPs, in particular, prescribe the conduct of all the specific individual activities that must be performed to support the objectives of the biosecurity plan. This includes, for example, procedures pertaining to: the receipt, storage, and inventory of BSATs, the inventory of animals or plants infected with high-consequence pathogens and toxins, housekeeping procedures for the containment laboratory, and the segregation and sterilization and/or destruction of potentially infectious materials prior to disposal. SOPs for afterhours access to the laboratory and implementation of the "two-person rule" (i.e., the requirement that at least two persons need to be present when BSATs are accessed or handled) should also be considered.

All SOP documents should be electronically accessible from a single centralized location. The first page of each SOP should clearly display the creation date and revision date for the document. The creation date corresponds to the date the SOP was originally created. The revision date corresponds to the date of the most recent change to the document. Ideally, SOP documents are under a process of revision control. Revision control not only prevents two individuals from inadvertently editing a given document at the same time but also ensures that multiple versions of the document are not simultaneously in use. Every time a staff member performs any task governed by a written SOP, this individual should be required to visit the centralized storage location for SOPs and obtain the latest version of the applicable SOP. This ensures that changes to procedures are instantly distributed across the entire organization and are, therefore, consistently implemented.

Training

The final component to ensure a successful biosecurity program and effective implementation of administrative controls is training. Training must be developed and delivered to all staff members regarding the biosecurity plan and all SOPs that have been developed in support of the biosecurity measures implemented by the organization. Initial training should begin with a review of biosecurity policies, their relationship to federal regulations and guidelines, and the purpose of the biosecurity program plan. The importance of communication must be emphasized. Properly executed training that incorporates adult learning techniques and measures competency (i.e., knowledge, skills, and abilities) should be delivered at least annually or when processes, procedures, or risks change. For maximum effectiveness, training should emphasize observable and measureable outcomes, which can be tracked by supervisors. Frequent recurrence of training activities is an effective way to rapidly ensure sustained compliance by staff. All staff members should be aware of the risks to the organization, the procedures for their mitigation, and their responsibilities within the biosecurity program.

Personnel Suitability and Reliability

Administrative components of operational biosecurity include the development of policies and procedures that allow access to high-consequence pathogens and toxins. This should be a stepwise process that begins with hiring. Human resources departments must be involved in the biosecurity process. It starts with the development of job descriptions for individuals who have a role to play in accessing, working with, or securing these agents. This includes security and safety staff, researchers, animal handlers, technicians, and facility maintenance and custodial staff. These individuals should undergo at a minimum background checks and suitability assessments (which involve security risk assessments). Some facilities may also require higher levels of security clearance and screening if the facility is performing government-funded defense research or other sensitive research activities. Post-offer pre-employment screening should occur, and the offer should be conditional on successful passage of these checks.

Some organizations also have policies on pre-employment drug testing. Generally, these are not as effective as random drug testing programs, but they are less costly in ongoing program administration. The exception is hair follicle testing. Generally, individuals who have used controlled substances can abstain for several weeks and have a negative urine test. Hair follicle testing captures a much longer window of use (usually several months or more) and is therefore much harder to defeat.

Material Control and Accountability

Programs and processes should be put into place that only allow access to high-consequence pathogens when employees have successfully completed documented training and demonstrated competency in compliance with all the organizational policies and procedures. In the U.S., only individuals who have successfully completed a security risk assessment conducted by the Federal Bureau of Investigation's Criminal Justice Information Services Division can access select agents or toxins.[7] Some organizations store these materials in locked freezers in limited access rooms. Whatever approach is used, it is generally more workable when fewer individuals have access. This becomes especially important when inventories are conducted. The chances of creating documentation errors when logging vials out for use increases significantly with the number of individuals involved in the process. Inventories must be conducted on a periodic basis. Each vial must be documented and accounted for, including its contents and physical location. This process can be achieved through low-tech solutions such as paper records and logs, or it can be accomplished through computer software programs specifically designed for this purpose. Whatever process is used, the end result must be an accurate record and accounting of each vial that exists in the system. Supplemental processes can also be utilized to save time. For example, stocks that are being saved for long-term storage (either because they are seed stocks or are being maintained as client study data) can be sealed in boxes with tamper–proof evidence tape after they have been counted.

Information Security

In addition to physical security measures there must be equal emphasis placed on the protection of the facility hardware and software. There must be firewalls and IT security measures to prevent outside entities from gaining access to data and sensitive electronic information. Such data and information could be used to override security measures, gain access to the facility, develop a greater and more detailed understanding of the research being conducted, and/or identify and/or locate the pathogens being used.

Material Transport

The biosecurity plan must address requirements for receipt, shipment, and storage of high-consequence pathogens and toxins as well as any other biological materials of value. Individual countries have specific requirements that should be incorporated into the plan. Processes and procedures must be developed to ensure that all high-consequence pathogens and toxins are received from or shipped to other entities that have authority to receive or distribute such agents. Steps must be taken to comply with all national and international transfer, shipping, and packaging regulations. In the U.S., all individuals who package or receive BSATs must successfully pass a security risk assessment. In addition, if the organism is a tier 1[8] agent (agents that pose the greatest risk), the individual must have a pre-access suitability assessment and an ongoing suitability assessment.[9] When shipments are received, they must be sent directly to the laboratory or temporarily stored in a secure location until they can be unpacked and added to the inventory. In addition, provisions must be developed for unexpected delivery of high-consequence pathogens and toxins. In some cases intratransfers (transfers of agents within the same organization) may occur. In this case, the biosecurity plan must identify the chain of custody and the documented procedures to prevent theft, loss, or release during transport.

Incident Response/Drills

An incident response plan must be developed by organizations handling, storing, or transporting BSATs. The plan must define response protocols in the event of a theft, loss, or release of high-consequence pathogens and toxins. Specifically, this plan must provide detailed procedures for emergency response team notification (intra- and extra-entity) in the event that an incident occurs. The incident response plan must take into consideration the insider and outside threats identified in the biosecurity risk assessment as well the potential for other incidents that could cause a potential release of high-consequence pathogens and toxins. These other incidents may be natural events such as hurricanes, earthquakes, or floods or intentional acts such as chemical or biological spills, fires, and workplace violence. The response plan should contain federal and local reporting requirements. There should be an accurate evaluation of the time it will take outside responders to reach the facility whether it's to interrupt a threat or to respond to an accidental release or spill. In the U.S., the new CDC regulations[10] require entities that possess tier 1 agents to ensure that a response team can be on-site within 15 minutes after receipt of an intrusion detection alarm. This has to be spelled out in the response plan and will require on-site security or local police to support the plan. In addition, the plan has to be tested and mock drills must be conducted to ensure that the time constraints can be met.

Management and Implementation

The biosecurity plan and all associated policies and SOPs should be reviewed immediately after any security incident and after all drills or exercises. If none of these events has occurred in the past year, the administrative controls should receive a comprehensive review at least annually. This allows the organization to confirm that previously identified threats still exist and that they remain properly prioritized. New threats emerge from time to time, and these, too, must be identified and appropriately prioritized for either management or mitigation. An annual assessment permits the identification of improved risk management strategies, new technologies for risk mitigation, and cost savings from reprioritized threats. Additionally, it also ensures that new staff members with responsibilities under the biosecurity program have an opportunity to be involved in the risk assessment process.

BIOSECURITY CHALLENGES ASSOCIATED WITH DIFFERENT INSTITUTIONS

The challenges associated with the implementation of an effective biosecurity program may vary depending on the organization and the amount of "institutional control" exercised by the senior leadership and mirrored in management practices of the organization. The definition of institutional control used in this context refers to the way an organization ensures compliance with external regulatory requirements and internal policies and procedures developed to minimize the organization's exposure to liability as a results of noncompliance with federal or state regulations. How the measures are implemented to ensure compliance may vary based on the type of organization where these pathogens are used or stored.

Academic Institutions

Institutions of higher learning such as colleges or universities present some of the most difficult challenges in the implementation of biosecurity control measures. This stems from many issues. The lack of what we will call "institutional control" is perhaps the most significant factor. Institutional control is management's awareness of the regulations and current violations combined with the willingness to take corrective action. There are also limited resources available in the academic setting to implement the measures needed to secure these pathogens. This is a result of the tenure track employment structure and academic freedom positions that many instructors have. The ability of a university biosafety officer or safety director to ensure compliance can be limited in a case where there are very few consequences for noncompliance. Tenure track employees cannot be harshly disciplined or discharged as easily as those working in the private sector. In addition, many academics believe they have the right to exercise their academic freedom and may resist adhering to regulations, which can add restrictions to their perceived ability to teach or conduct research.[11] Even if there are no overt intentions of wrongdoing, oversight tends to be more relaxed when a culture of compliance has not been required as a condition of employment. As a result, some of the more serious events of noncompliance with the CDC select agent regulations contained in 42 CFR Part 73 have occurred in academic environments.[12]

As stated in BMBL, "The success of both [biosafety and biosecurity] programs hinges on a laboratory culture that understands and accepts the rationale for biosafety and biosecurity programs and the corresponding management oversight."[13] In addition, there is the need for continuous education and enforcement as a result of the high turnover rates of individuals who may have access to high-consequence pathogens. Students who may have been trained appropriately eventually graduate and move out of the research program. They are replaced by new students who need training. This is a constant challenge for the safety departments and biosafety officers, who already have limited resources.

Government Facilities

Some of the same issues found in academic institutions can be found in government facilities as well. Historically, some of the worst biosafety and biosecurity practices have been found in government laboratories.[14] Although government facilities may have the resources and access to security forces to support the implementation of physical security measures, many times accountability and institutional control are lacking. This may be due in part to the difficulty in disciplining civil servants but may also be a result of the historically relaxed approach to compliance that government entities have had toward their own regulations. Typically, government regulatory agencies have limited ability to penalize other agencies for noncompliance, and the federal government is typically indemnified from lawsuits due to sovereign immunity. Government entities need to be held to the same or higher standards as private entities. Often government facilities have large libraries of reference organisms with the same threats and vulnerabilities as private entities. This makes them a potentially more appealing target for outsider and insider attacks.

Private Organizations

Private and commercial entities generally have the greatest incentive and ability to ensure the protection of pathogens and toxins from misuse and to ensure compliance with regulations that require measures to improve biosecurity. This is generally due to the fact that they have historically been held to stricter standards from the Occupational Safety and Health Administration, Environmental Protection Agency, Food and Drug Administration, and other regulatory agencies. Consequently, a culture of compliance has been firmly established and maintained. Companies that fail to comply with federal regulations suffer consequences such as product liabilities, employee or public lawsuits, and an inability to receive federal funding or private contracts for research. Pressure from shareholders or board members is also a factor if employees have been cited or fined for noncompliance. And, finally, it is generally much easier to hold private employees accountable for adherence to company policies and procedures, provided that they have been trained appropriately and the requirements are documented.

NEW REGULATORY REQUIREMENTS

On October 5, 2012, the CDC and Animal Plant Health Inspection Services (APHIS) published new regulations in 7 CFR Part 331, 9 CFR Part 121 and 42 CFR Part 73 entitled Possession, Use, and Transfer of Select Agents and Toxins; Biennial Review. With APHIS and CDC approval, entities are now registered for a maximum of three years as opposed to the current five years.

BSATs are now tiered to clearly distinguish those materials with the greatest potential for deliberate misuse that could result in devastating effects to the economy, critical infrastructure, and public confidence. The agents and toxins that present the greatest risk are in tier 1 and include: Ebola virus, *Francisella tularensis,* Marburg virus, Variola major virus, Variola minor virus, *Yersinia pestis,* Botulinum neurotoxin, and Botulinum neurotoxin-producing species of *Clostridium, Bacillus anthracis, Burkholderia mallei,* and *Burkholderia pseudomallei.*

Organizations in possession of tier 1 BSATs are now required to specifically and clearly address certain physical security, electronic security, and material control and accountability issues within their security plan. The security plan must be submitted to CDC/APHIS for review along with other documentation at three points in time: when the entity initially registers for possession of select agents and toxins, upon each renewal of registration, and at any other time, upon request.

The security plan must include policies and procedures that describe how initial suitability assessments and ongoing suitability monitoring will be performed by the organization for persons who will or do have access to tier 1 materials. Staff members with access to these materials are now required to receive training in these policies and procedures. The security plan must also detail the process by which individuals will be granted access to tier 1 materials. The new regulations also require that all employees be trained in the policies and procedures for reporting, evaluation, and corrective actions concerning personnel suitability assessments. Organizations must now develop and document a mechanism to report incidents or conditions that could affect an individual's ability to work safely with BSATs or to safeguard them from theft, loss, or release. Finally, the security plan must describe procedures by which the responsible official will coordinate his or her efforts with the entity's safety and security professionals to ensure security of tier 1 select agents and toxins and share relevant information, as appropriate.

Entities with tier 1 select agents and toxins are now required to have a minimum of three security barriers, which must be monitored 24 hours a day, seven days a week, to detect intentional and unintentional circumvention. Each barrier should increase the time of delay for an individual to reach the secured area where select agents and toxins are used or stored. The final barrier must be protected by an intrusion detection system (IDS) unless it is physically occupied. The organization must be staffed with personnel who actively monitor the IDS, who are capable of alarm evaluation and interpretation, and who can alert the designated security response or law enforcement force. Response time is measured from the time of an intrusion alarm or report of a security incident to the arrival of the responders at the first security barrier.

Entities are now required to conduct complete inventory audits of all select agents and toxins in long-term storage under the following circumstances:

- Upon the physical relocation of a collection or inventory
- Upon the departure or arrival of a principal investigator
- In the event of a theft or loss

There is a cost to the organization associated with the time required to conduct the required audit(s), especially when there must be a procedure in place to protect the integrity of the research associated with the inventory.

In addition, procedures must be developed so that access to the laboratory outside of normal business hours is limited strictly to those individuals specifically approved by the responsible official or designee. The biosecurity risk assessment must specifically include procedures to screen visitors, along with their property and vehicles, at the entry and exit points of the registered space, or at other designated points of entry to the building or facility.

Finally, additional procedures must be established to provide backup security measures in the event that access control systems, surveillance devices, and/or systems that manage the requirements of the final rules are rendered inoperable (the use of security guards, for example). Staff members must now be provided with security awareness training as well as incident response training.

New requirements for material control and accountability include a provision to safeguard animals or plants intentionally or accidentally exposed to or infected with a BSAT. The associated processes must be described within the entity's security plan, biosafety plan, and incident response plan. Furthermore, entities must now put in place consistent practices for the shipment, receipt, and storage of BSATs, including documented processes for the security and tracking of the shipment after disposition.

Facilities possessing tier 1 BSATs must conform with new electronic (information technology) security requirements. This includes the security of networks, computers, peripheral devices, and data storage devices that possess select agent and toxin information, files, and applications. Only authorized and authenticated users will be granted access to this information and the equipment on which it is stored or manipulated. Access is limited to the extent necessary for users to fulfill their roles and responsibilities. Further, a user's access will be modified when her roles and responsibilities change or when access to BSATs is suspended or revoked.

The entity must ensure that adequate controls are in place to prevent malicious code (including but not limited to computer viruses, worms, and spyware) that may compromise the confidentiality, integrity, or availability of information systems that manage access to registered spaces. Defined processes are now required to ensure that updates and regular patches occur to protect the integrity of the operating system.

All external connections to the electronic systems that manage security for the registered space must now be isolated or possess controls that permit access only to authorized and authenticated users and that monitor their electronic activities.

CONCLUSION

Operational biosecurity encompasses all components that must be considered and implemented to ensure that a comprehensive approach has been developed to address all aspects of biosecurity in the work environment. These include administrative controls, physical security, and human resource components that individually and collectively play a part in the security of high-consequence pathogens and toxins and valuable biological materials.

References

1. Centers for Disease Control and Prevention and the National Institutes of Health, Biosafety in Microbiological and Biomedical Laboratories, 5th edition p. 105 (2009).
2. Select agents as defined by CDC and APHIS 42 C.F.R. Part 73, 7 C.F.R. Part 331, and 9 C.F.R. Part 121.
3. http://www.ilpi.com/msds/ref/administrativecontrols.html (accessed 14 December 2012).
4. http://www.cdc.gov/biosafety/publications/bmbl5/ (accessed 14 December 2012).
5. ftp://ftp.cenorm.be/public/CWAs/wokrshop31/CWA15793.pdf (accessed 14 December 2012).
6. http://www.cen.eu/cen/pages/default.aspx (accessed 14 December 2012).
7. Guidance for Suitability Assessments; Centers for Disease Control and Prevention (CDC), Division of Select Agents and Toxins and the Animal and Plant Health Inspection Service (APHIS), Agriculture Select Agent Program (1 October 2012). http://www.selectagents.gov/resources/Tier_1_Suitability_Guidance_version_2_1.pdf (accessed 14 December 2012).
8. Tier 1 list includes Ebola virus, *Francisella tularensis,* Marburg virus, variola major and minor viruses (smallpox), *Yersinia pestis* (plague), *Clostridium botulinum* and botulinum toxin, *Bacillus anthracis* (anthrax), *Burkholderia mallei* and *Burkholderia pseudomallei* (melioidosis).
9. Security Guidance for Select Agent or Toxin Facilities; Centers for Disease Control and Prevention, p. 21 (October 1, 2012). http://www.selectagents.gov/resources/Security_Guidance_version_2_English.pdf (accessed 14 December 2012).

10. Health and Human Services, 42 CFR Part 73; Possession, Use and Transfer of Select Agents and Toxins, Biennial Review; Final Rule (5 October 2012).
11. 1940 Statement of Principles on Academic Freedom and Tenure. http://www.aaup.org/nr/rdonlyres/ebb1b330-33d3-4a51-b534-cee0c7a90dab/0/1940statementofprinciplesonacademicfreedomandtenure.pdf (accessed 14 December 2012).
12. Couzin, Jennifer. Lapses in Biosafety Spark Concern. *Science.* Vol. 317, no. 5844, p. 1487 (2007).
14. Centers for Disease Control and Prevention and the National Institutes of Health. Biosafety in Microbiological and Biomedical Laboratories, 5th edition, p. 106 (2009).
15. Personal communication, USAMRIID documented incident.

CHAPTER 6

Basic Principles of Threat Assessment

Benjamin A. Perman, Lindsay T. Odell, Jason Griffeth, and Daniel N. Apple

INTRODUCTION

Understanding potential threats to facilities and programs using dangerous biological agents or valuable biological materials (VBM) is necessary to building viable and sustainable biosecurity programs. Insider threats are staged in a series of deliberate acts. Understanding the behavior leading to the implementation of a biological weapons attack or deliberate theft or diversion of VBM may be the most powerful tool in detecting the adversary, interrupting his planning and actions, and ultimately removing or limiting the threat.

A COMPREHENSIVE APPROACH TO BIOSECURITY

Preventing bioterrorism begins with preventing biological materials, information, and expertise from being lost, stolen, or somehow diverted by terrorists or their accomplices. Biological materials and the information necessary to misuse them for criminal purposes are unfortunately easily transported and transferred. There are few if any technical countermeasures that can either detect them quickly enough to interdict criminals or physically prevent their removal from designated and otherwise safe laboratories. Biological materials, particularly infectious organisms, can be recovered from the environment, animals, and infected humans. As discussed in previous chapters, certain inherent characteristics (i.e., microscopic size, ability to reproduce) of biological materials pose considerable challenges to effective biosecurity programs and policies. Even if a limited amount of material can be obtained by a prospective terrorist, many infectious organisms can be easily grown *in vitro* to yield large quantities. Most biological materials can be easily obtained, transferred, and grown and are therefore particularly vulnerable to persons who are inside the biomedical research, public health, or clinical communities. Biosecurity often tends to be a matter of insider threat mitigation as security programs are forced to focus on inherently human behavioral factors to mitigate the risks that biological materials fall into the possession of terrorists or other criminals.[1,2,3] As a result, biomedical research, clinical, and diagnostic laboratories along with government policymakers are turning to insider threat mitigation strategies. Insider threats are posed through a series of deliberate acts. Human behavior is perhaps the most powerful tool in detecting the adversary, and interrupting planning and actions is the most effective mechanism for ultimately removing or limiting the threat.[4] This approach is very well suited to biomedical research, public health, and clinical communities and represents a solution to the central problems in biosecurity. This approach also demonstrates promise in countering the threat of bioterrorism while at the same time protecting the integrity, safety, and security of research and healthcare institutions.

The field of biosecurity, and even the term itself, has generated considerable confusion and merits some clarification. In this chapter, biosecurity refers broadly to the entire spectrum of security around research or clinical laboratories, public health, and veterinary and clinical institutions. The ultimate goal of biosecurity is to protect valuable biological materials and related sensitive information, clinical and research staff and, most importantly, the integrity of host institutions' missions and the biomedical community itself.[5] A holistic approach to security is usually the best. Taking a comprehensive approach to biosecurity is particularly helpful when considering the concept of threat, because it is harmful to the implementation of effective security programs to separate physical security measures from threat mitigation tactics. This is particularly relevant in the biomedical and microbiological sector, where security programs must be successfully integrated into other highly specialized safety and scientific systems. Biosecurity is a system of measures and practices designed to counter the loss of control or release of infectious agents, toxins, or related sensitive information from a designated laboratory or clinic. A biosecurity program has multiple components: physical security controls, procedural controls, administrative controls, and training.

SAFETY VERSUS SECURITY: RISK VERSUS THREAT

As discussed in previous chapters, it is important to carefully define related terms that have meaning in biosafety or biomedical science in order to avoid confusion. The commonly used language around the concept of risk is an important area that often causes a great deal of confusion when biosecurity is discussed in a biosafety or scientific setting. It is important to distinguish biosecurity risk from biosafety risk. Biosecurity risks are determined by measures of vulnerability to and consequences of deliberate acts, as well as the dynamic threats that impact the system or its assets. Biosafety risks deal with probabilities and consequences of accidents where no nefarious cause exists. Vulnerabilities are individual or systemic weaknesses that can potentially be exploited by adversaries to defeat a security system.[6] Individual vulnerabilities can aggregate so that the defeat of a single critical security measure or series of linked or interdependent security measures can lead to the defeat of an entire security system. These systematic vulnerabilities usually render security systems ineffective.

Both forms of risk (safety and security) are, in different ways, tied to human factors; however, in biosecurity there are conscious actions, often including a planning phase, that precede the events that are being measured or estimated in any assessment of security risk. The preceding indicators are the subject of the threat assessment. Threats in security, and indeed in a biosecurity context, directly involve human factors. It is important to consider that threats can exist both inside and outside a security perimeter and can be an internal, external, or an indirect threat. All threats are either overt (announcing their presence and intent) or clandestine. Whereas many risk assessments include a purely quantitative assessment, threat assessments cannot be adequately described in numerical terms.

Adversaries

The concept of an adversary is central to understanding a threat-based approach to security. Fundamentally, adversaries are people. An adversary may be a single individual or a group of individuals who have the intent to cause harm to a facility, its personnel, or its function. Unlike the term "threat," the term "adversary" refers specifically to an individual or a number of individuals, sometimes called "actors" or "threat actors." The specific motivations of adversaries can vary a great deal across a continuum of diverse objectives, including but certainly not limited to disruption of research, theft/diversion of research or materials, destruction of materials, and targeting of persons for reasons unrelated to an institute's scientific or healthcare mission. Because adversaries are persons, it is possible to learn a great deal about them from case studies in the biomedical and microbiological field in an effort to establish trends in behavior and tactics. Analyzing case studies allows us to better understand how future adversaries will operate; indeed, this is how the discipline of threat assessment was developed.[7] The National Institute of Justice commissioned a study of known "attacks on principals" to understand the indicators that predicted violent action in adversaries that posed a direct threat to public officials, celebrities, and significant leadership figures in private industry. The results of this study formed the Exceptional Case Study, an ongoing effort within the U.S. Secret Service that continually refines our understanding of adversaries and the behavioral indicators that predict threatening behaviors, including violent acts themselves.[7,8] The results of the Exceptional Case Study are applicable to understanding security risk in a life sciences context, as the behavioral indicators that predict threatening behaviors and risks of violence are the same behavioral indicators that can assist us in identifying both insider and outsider threats within a laboratory environment.[9]

Threats in the biological sciences follow a similar trajectory in the evolution of threatening or violent behavior. These similarities support the use of traditional behavioral indicators in developing a threat assessment approach tailored to specific biosecurity requirements.[9] In a preliminary analysis of cases involving known adversaries who operate in the biomedical sector we have represented each adversary type at least once in a preliminary sample. The analysis could not be limited solely to laboratories working with current U.S. biological select agents and toxins (BSAT) because all types of threats are not represented in cases at these institutions. To cover all threat types, we have broadened the scope of the analysis to include threats at a variety of biomedical institutions (including public health and clinical/veterinary medicine), regardless of specific motives, to model the targeted violence process in this community. The preliminary analysis tends to show that threats, particularly insider threats, evolve along very similar pathways in this community. For example, threats involved in animal rights extremism, anti-abortion/anti-stem cell extremism, workplace violence, research misconduct, stalking, and select agent malfeasance in the biomedical community all seem to have significant overlapping threat behavioral characteristics in what would otherwise seem to be very different scenarios. From this perspective we can apply established best practices in threat assessment to the insider threat mitigation challenges in biosecurity. Table 6.1 provides examples of individual or group adversaries.

Threats

Unlike adversaries, who are individuals or groups without a targeted intention or developed capability, threats represent a larger concept because they include a measured understanding of the capabilities and intentions of the

TABLE 6.1 Adversaries with selected examples

Individual Adversaries	Examples of Individual Adversaries
Employees with access	Bruce Ivins[a] Diane Thompson[b] Raymond Clark[c]
Employees without direct access	Deborah Green[d] Richard J. Schmidt[e] Brian Stuart[f] Amy Bishop[g]
Students and other affiliates	Seung-Hui Cho[h] Aafia Siddiqui[i]

[a] Dr. Bruce Ivins remains the primary suspect in the FBI's Amerithrax investigation. Ivins is implicated in the theft *of Bacillus anthracis* from the U.S. Army Medical Research Institute for Infectious Disease (USAMRIID) and subsequent attacks involving the mailing of envelopes containing *B. anthracis* spores to media and U.S. government personnel (DoJ, 2011). The case against Ivins rested on an extensive threat assessment that included clinical psychiatric/ psychological analysis.[18-20]

[b] Diane Thompson, a technician in the clinical diagnostic laboratory of the St. Paul Medical Center Hospital in Dallas, TX, deliberately infected 12 people with *Shigella dysenteriae* type 2.[21]

[c] Raymond Clark, a vivarium veterinary technician, pled guilty to the murder and sexual assault of Yale University graduate student Annie Le in 2009. Although no motive was given, one report indicated that Clark was officious when it came to rodent handling rules and complained to supervisors and research staff about mishandling animals.[22]

[d] Dr. Debora Green, a former oncologist, was convicted of attempted murder charges based on a claim that she had tried to use ricin to poison her estranged husband, Dr. Michael Farrar. Green was also convicted of murdering two of her three children by arson.[23]

[e] Richard J. Schmidt was a married Louisiana gastroenterologist who had resisted efforts by his victim to end a 10-year amorous relationship. He routinely gave the victim vitamin B-12 injections and used an HIV-contaminated hypodermic, later linked to one of his patients, to deliberately infect her.[21] Schmidt was convicted of second degree murder.

[f] Brian Stuart, a Missouri phlebotomist, was convicted of first degree assault for deliberately infecting his 11 month-old son using a blood sample obtained from the hospital he worked for. The incident occurred during a paternity dispute with the boy's mother.[21]

[g] Amy Bishop was an assistant professor of biology at the University of Alabama, Huntsville, who murdered three other biology professors, apparently in retribution for denying her tenure in 2009. Bishop had a history of violence that began with the 1986 shotgun death of her brother in Massachusetts.[24]

[h] Seung-Hui Cho is responsible for the single most violent active shooter case in U.S. history, killing 32 students and wounding an additional 17 at Virginia Tech University in 2007.[16]

[i] Aafia Siddiqui, a Pakastani woman, was arrested in Afghanistan and convicted of attempted murder of U.S. forces. At the time of her arrest she was carrying a USB storage device that contained protocols for mass casualty weapons and details of potential attack targets including New York City landmarks and the Plum Island Animal Disease Center. She was also carrying approximately one pound of sodium cyanide, according to the indictment in the U.S. District Court of Southern New York. Siddiqui had received a Ph.D. in neuroscience from Brandeis University while living in the U.S. Her second husband is the nephew of detained Al Queda planner Kahlid Sheikh Mohammed, who referred several times to Siddiqui's actions in interviews after his arrest in Pakistan.[19]

(Table continued on next page)

TABLE 6.1 *(Continued)*

Individual Adversaries	**Examples of Individual Adversaries**
Visitors	Eric Robert Rudolph[j]
General Criminals	Numerous
Lone Wolf (ideological)	Larry Ford[k]
Limited Objective/Petty Larceny "Drugstore Cowboy"	Numerous
Group Adversaries	**Examples of Group Adversaries**
Ideological Domestic Extremists/Terrorists[l,m]	Rajneesh Cult Army of God Stop Huntington Animal Cruelty (SHAC) Animal Liberation Front (ALF)
Ideological International Terrorists	Al Qaeda on the Arabian Peninsula, Hamas
Foreign Powers	South African (Apartheid) Soviet Union Iraq Iran Libya North Korea Syria

[j] Eric Robert Rudolph was arrested on suspicion of bombing multiple abortion clinics and sites at the Atlanta Olympics. In several cases he gained initial access to the clinics as a visitor, befriending administrative staff to obtain information on clinic operations and security.[23]

[k] Dr. Larry Ford is perhaps one of the most enigmatic of these adversaries. Ford was the primary suspect in the 2000 attempted murder of his business partner at an Irvine, CA, biotech company engaged in developing a female suppository contraceptive.[25] In the course of the police investigation several aspects of Ford's background came to light, including white supremacist writings, a cache of illegal munitions at his home, mysterious storage of *Clostridium* species, accusations of attempts to deliberately infect six mistresses, ties to the Surgeon General of the South African Defense Force, and claims that he was employed by the CIA.[25,26]

[l] Diane Yvonne Onang (a.k.a. Ma Anand Puja), a nurse practitioner and secretary-treasurer of the Rajneesh Medical Corporation, used her position and title to successfully obtain *Salmonella typhimurium* from American Type Tissue Collection. Salmonella was used to contaminate salad bars at restaurants in The Dales, OR, to adversely affect voter turnout in a local election.[23]

[m] James Holmes is suspected of the July 20, 2012 killing of 12 and injuring 58 persons in the Century 16 movie theatre in Aurora, CO. Holmes had recently left a Neurobiology Ph.D. program and the University of Colorado, Anschutz. During the months prior to the massacre, Holmes acquired four firearms and over 6,000 rounds of ammunition, tear gas cylinders, explosives, personal protective equipment and set up an elaborate booby-trap at his apartment in another part of Aurora. In the weeks prior to the attack Holmes made 3 separate reconnaissance trips to the movie theatre to study the doors and other security and physical features of the interior of the theater (Ingold and Fender, 2013).

adversary. It is important to consider that threats can exist both inside and outside a security perimeter, as is seen in the three different types of threat:

1. Internal threat
2. External threat
3. Indirect threat

In addition, threats may be overt (visible or self-announcing) or clandestine. Clandestine threats present an obvious detection and assessment challenge that can be addressed by the skilled application of threat assessment practices. Internal threat factors are direct threats that involve employees and visitors with authorized access to an institution. Security and human resources managers often attempt to manage internal threat through administrative controls and codes of conduct; however, a threat assessment provides a means to detect behaviors of concern and initiate mitigation measures when threats do not comply with administrative controls. External threat involve all direct threats to an institution or individual from outsiders who do not have formal access to the institution. Finally, indirect threats are those that result from an intermediary, sometimes referred to as "cutout," or from the proximity of unrelated criminal or violent activity to individuals or institutions. Specific examples that further elucidate the three threats are provided in Table 6.2.

The Targeted Violence Process

Threat assessment is a discipline that has developed to support security managers and planners who are facing a threat that cannot easily be countered through traditional physical security measures alone. It is used extensively in the arena of personal protection, also known as executive protection, where physical perimeters and access controls are not present or where insiders pose a significant potential threat.[10] The latter is also a primary concern in biosecurity. Threat assessments are used to evaluate the potential for hostile actions based on communicated, demonstrated, or perceived events. Threat assessments take into consideration internal, external, and indirect threats and use a formalized analysis called the targeted violence process. This process describes three distinct phases (intent, ability, and opportunity) to better understand the progression of threat behaviors. The threatening behaviors themselves are usually the primary study point within an assessment. It is important to emphasize that threatening behavior is very different from eccentric behavior, and not all improper communication or action is inherently threatening. Assumptions that threat assessment professionals consider eccentric behavior to be threatening confuse the process of threat assessment and alienate threat

TABLE 6.2 Internal, external and indirect threats

Internal Threats
Employee making direct threats against employer: internal sabotage or work place violence
Employee making direct threats against research: internal sabotage
Employee involved in threatening behaviors or subversive activities: social networking
Internal animosity or issues leading to destabilization of security protocols: love triangles across resource boundaries, destructive relationships between authority lines
External Threats
Direct threats against employees (personal connection or terrorist related)
Direct threats against the employer or supervisor (work place violence or terrorist related)
Direct threat against the research or institute (external sabotage or terrorist related)
Indirect Threats
"In the line of fire" or "wrong place at wrong time" (work place violence or terrorist related)
Inadvertent release of information (general criminal activity that leads to external or internal factors)
Elicitation of information or manipulation by third parties

assessment professionals. The alienation of the threat assessment professional in an insider threat mitigation program puts critical employee participation at risk, particularly peer- and self-reporting functions, which are discussed later.[7]

The fundamental concept of the targeted violence process is the recognition that threats will evolve along a concerted path, divided into three developmental phases, intent, ability, and opportunity, that are defined by the threat's potential violent action. Fein et al. (1995) noted that "Violence is a process as well as an act. Violent behavior does not occur in a vacuum. Careful analysis of violent incidents shows that violent acts often are the culmination of long-developing, identifiable trails of problems, conflicts, disputes, and failures."[10] In the targeted violence process, the three developmental phases (intent, ability, and opportunity) are in turn divided into subcategories that determine the severity of the threat posed: grievance, ideation, planning, preparation, and implementation. The targeted violence process is shown as a schematic in Figure 6.1. This figure is showing the progression through each phase, intent, ability, and opportunities, and the related subcategories. It also depicts the impact of preventive and responsive measures, showing the clear benefit of preemptive measures such as suitability programs. Proper identification of the subcategories can assist the threat assessment professional in making a sound determination regarding the current threat and the level of intervention and mitigation needed.

The phase in which the actor has decided on a course of action and begins to prepare to carry it out represents the first developmental category in the targeted violence process, intent. Intent includes two subcategories of action: grievance, which is a complaint generated by a single precipitating event or series of events leading to the concept of a perceived wrong, and ideation, which is the development of a course of action to remedy the perceived wrong. The remedy may not be rational and may not appear to be a remedy to anyone other than the adversary experiencing the grievance and ideation. Additionally, the phase of intent may be manufactured: an individual or employee may be acting under duress or coercion from and in collusion with an adversary. Here the grievance and ideation rest in the outsider, who relies on the insider to fulfill the remaining steps in the evolution of the targeted violence process. Ability is the phase that involves the actor's effort to develop the framework for action. Ability has two subcategories of action: planning, which involves the development of the action to be carried out; and preparation, the gathering and staging

FIGURE 6.1 A schematic depiction of the targeted violence process and the role threat assessment to interrupt it

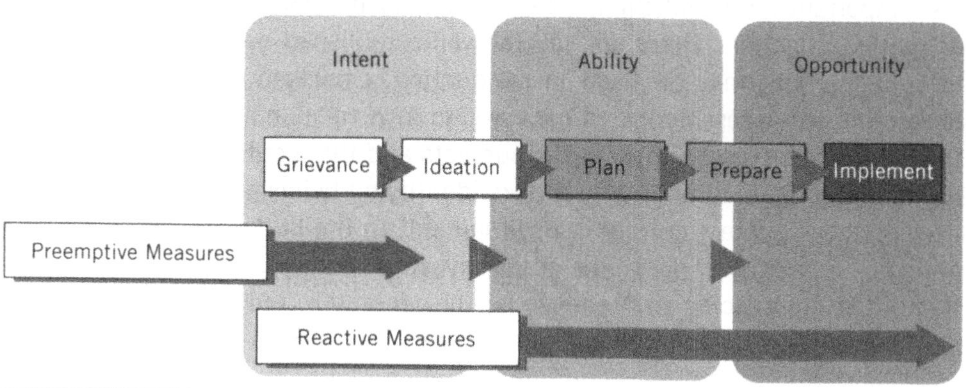

of the tools and plans to carry out the action. Opportunity is the phase in which the actor has met all needed criteria to carry out an action; it includes only one sub-category, implementation, or simply the act of carrying out of the planned action. Opportunity is the intersection of the targeted violence process, or the human component, and vulnerabilities within the system that the actor exploits to implement a hostile act. Important to note is that the hostile act is not limited to violence within a biosecurity context but can include theft or diversion of VBM and related sensitive information.

Preemptive Measures

Threat assessment can be thought of as an approach that seeks to recognize threats, characterize their development, and interrupt the targeted violence process. As such there are three major functions of a threat assessment program:

1. Identification of a potential perpetrator
2. Assessment of the risks of violence or other hostile action posed by a given perpetrator at a given time
3. Management of both the subject and the risks that he or she presents to a given target

The threat assessment measures that are employed can be preemptive or responsive. There are several preemptive measures that can be applied in threat assessment. Preemptive measures, if properly employed, can be used to identify threats and interrupt the targeted violence process early in its development. Generally, preemptive measures are various forms of background investigations or suitability screening processes that can be used to identify behavioral indicators in a subject's background. Background or suitability investigations are also important for establishing a behavioral baseline for the individual subject of the investigation. There are several well-established examples of available information that may be used in conducting a background investigation to develop threat assessments.[7] A background investigation should be tailored to the environment and culture of the subjects and the institute that is conducting the investigation; otherwise it risks identifying behaviors or trends that are normal in the subject culture or environment. In the biomedical research and clinical communities, the scope of the investigation may extend to cover other items that are specific to the basic science research, public health, pharmaceutical, veterinary, and clinical communities (Table 6.3).

TABLE 6.3 The scope of suitability investigations

Traditional Protective Investigation Scope
Titles
Liens
Residences
Licenses
Passport and visas
Travel
Receipts
Bank records
Credit
Neighbor interview
Friend interview
Coworker interview
Work associate interview
Weapons
Arrests
Warrants
Convictions
Court orders
Military records
Mental health findings
Pre-employment Screening and Credentialing for Biomedical Research, Public Health and Clinical Staff
Verification of educational credentials
Recognizable and credible references in the field of endeavor
Insured against malpractice
Possess applicable state licenses, registrations
Members of appropriate fellowships
A member of appropriate professional societies and credentials
Board certifications and examinations
DEA number
Publications
Other abstracts
Patents
Select agent program security risk assessment

Responsive Measures

Responsive measures seek to interrupt the targeted violence process once the threat is identified. Responsive measures are not by any means less effective than their preemptive counterparts; rather, they are necessary due to the fact that threats do not always develop along a concerted or sequential pathway. It is possible that a threat may first be identified and characterized in the ability and opportunity phases; therefore, threat assessment requires that several mitigation capabilities be available in the form of comprehensive response or mitigation plans. The specific mitigation plan should be calibrated to the threat's current stage in the targeted violence process.

Mitigating Intent

In the earliest stages of the targeted violence process the challenge is to develop tangible plans to mitigate intent. Successful mitigation with little or no law enforcement (LE) or outside professional assistance is possible but requires a highly diligent, proactive security approach that is integrated into all aspects of institutional leadership, including human resources, supervisors, and all facets of work-life support resources that are available to the subject. Mitigating intent also places a premium on peer- and self-reporting mechanisms to make initial behavioral indicators known to the subject's colleagues and provide a safe venue and procedures for reporting without fear of attribution or reprisal. Basic personal security training is also required so that colleagues and potential subjects (in a self-reporting scenario) know specifically what types of behaviors or circumstances to report. The threat management component of mitigating intent is not dependent on simply placating the subject, however; it can involve conflict resolution and institutional methods for redress of grievances. It is important that all parties act quickly and decisively to effect mitigation at this developmental phase and avoid escalation or further development of the threat within the targeted violence process.

Mitigating Ability

Although detection of threats that are in the ability phase of threat development may be easier, successful mitigation internally is more difficult because law enforcement resources or some form of intervention by a properly trained physical security or mental health professional is likely to be required. As was implied in the previous discussion of the subcategories of ability (planning and preparation), the threat may now be capable of violence. Worse, the threat may have already planned a response or counteraction to mitigation or intervention efforts. The risk of danger should never be taken lightly. The threat management strategies and resources employed to mitigate intent may not work and, if employed, may expose persons to undue risk.

Mitigating Opportunity

Once the threat is in the opportunity phase, both the security plan and the assessment process have essentially failed and the mitigation options become reactive. These purely reactive mitigation options, however, are least likely to result in successful threat mitigation. Mitigation requires a rapid law enforcement response that hopefully can be mounted before the threat has time to act or carry out an act if already in progress. If the responding law enforcement agency is being contacted for the first time, officers must be briefed on the nature of the threat so they can plan their actions appropriately to account for all safety provisions. The immediate threat of violence in a microbiological or biomedical setting has very unique officer safety requirements. It is prudent and, in the BSAT regulatory framework, required for an institute to have and exercise annually a comprehensive response plan. The response plan should consider mitigation of threats in the opportunity phase of their development.

Physical Security

There is a tendency in the application of physical security efforts to maintain a reactive state, a constant state of readiness to respond to a spectrum of security situations or contingencies. As a result of this tendency, security policies and procedures are generally written from a symmetric standpoint in which all vulnerabilities are identified and mitigated or countered within the cost and resource constraints of the institution. Most security professionals themselves are trained to "observe and report," not to analyze and mitigate. Therefore, information collection and analysis is underutilized, resulting in the need to cover all contingencies and vulnerabilities rather than prioritizing physical security solutions that address the specific targets that the threat has planned and prepared for. By not making specific preparations for the threat, reactive security systems are spread thinly across all vulnerabilities. As a result, reactive plans are almost always resource-constrained and can only successfully mitigate manageable situations such as fires or maintenance issues. Reactive-state security systems, however, fall well short when dealing with asymmetrical evolutionary situations that involve a determined, intelligent adversary, particularly one who has been allowed to prepare and practice an attack. Such adversaries are capable of modifying their tactics to account for and overcome static physical security measures but not without executing actions and behaviors that can be successfully detected and assessed by proactive measures.

Law enforcement agencies tend to be, largely, reactive entities. Until a crime is committed, law enforcement assets are very limited in their ability to engage or remove the threat. With the exception of some agencies that have developed prevention strategies in response to domestic terrorism, stalking threats, and campus violence threats, most law enforcement agencies are not effective at single-handedly countering determined threats. The modern concept of community policing and patrol, which calls for dispersing and moving deployments rather than targeted tactical approaches, put police forces themselves in a reactive state.

Physical security in a traditional sense overlooks the concept of threat assessment, and most standard physical security plans are based purely on a reactive model. Because threat assessment is a proactive measure that draws on unique capabilities in collecting and analyzing information, it is usually not included in these standardized plans. Most standardized plans fall short when dealing with an evolving threat situation. A proactive state takes advantage of information derived from threat assessment to help security managers prepare active countermeasures to prevent security situations. Countermeasures in the proactive state are very different from those in a reactive state. Proactive countermeasures are commonly deployed in proactive executive protection or close-protection settings and can include real-time analysis, surveillance/countersurveillance, avoidance, and tactical planning of reactions to various contingencies. A proactive state characterizes close protection because there are none of the luxuries of perimeter security and large, standoff distances. Only with input from threat assessment can security managers properly address threats and achieve more targeted security solutions. This enhanced proactive state creates threat-driven physical security.

THE PURPOSES AND REQUIREMENTS OF BASIC SUITABILITY AND THREAT ASSESSMENT PROGRAMS

Suitability and threat assessment programs are distinct elements of a personnel resource management system, which may be part of or combined with a personal reliability program, often referred to as "biosurety" (see Chapter 1). Suitability programs seek to establish an initial behavioral baseline that can then be periodically assessed in light of specific behavioral indicators. An additional measure that can be added to a biosecurity program is a threat assessment program. Threat assessment programs are focused on identifying potential perpetrators by assessing the risks of violence posed at a given time and managing both the subject and the risks that he, she, or they present to a given target. Each of these elements, suitability, reliability, and threat assessment, can be used as part of a laboratory biosecurity management program. An expanded description of suitability and threat assessment programs in laboratory biosecurity management is provided below.

Suitability Programs

Suitability is an assessment of the ability of an individual to work constructively within a given work environment. It is the determination of whether or not a person has the right credentials, skills, and disposition to be productive, safe, and secure in the work environment, both individually and as a part of a team. Implementation of suitability as part of employee screening prevents inappropriate allocation of human resources, which can erode morale, waste resources, and create an unsafe work environment, by ensuring that individuals are neither over- or under-qualified for their positions. Suitability programs help to ensure that employees are in the roles that fit them best and provide protection to the institution and fellow practitioners against fraud and imposters.

The U.S. government has identified suitability as an important component in a biosecurity program to prevent the deliberate or accidental loss or misuse of especially dangerous biological agents. The National Science Advisory Board for Biosecurity (NSABB), the National Academies of Science, and the Federal Experts Security Advisory Panel (FESAP) called for revision of the U.S. select agent regulations. These experts have all identified personnel security vulnerabilities and recommended some form of reliability or suitability program to support insider threat mitigation.[1,3,11] This chapter will not attempt to differentiate between reliability and suitability in the interest of clarity and will focus on the concept of suitability. At the time of writing the select agent programs (CDC and APHIS) have released a Final Rule that includes elements of a mandatory suitability program for entities housing tier 1 select agents (7 CFR 331, 9 CFR 121, 42 CFR 73).[13,14] In addition, elements of suitability programs are increasingly being implemented at U.S. institutions that sponsor select agent and other infectious disease research and have been described in the Biosafety in Microbiological and Biomedical Laboratories (BMBL) publication.[12]

Some elements of a typical suitability program include a suitability investigation to establish a behavioral baseline or to determine whether a candidate is fit for the position. Fitness can be described in a number of ways, ranging from being physically qualified to participate in a special immunization program, psychologically fit for work in confined spaces, or having no derogatory information on a criminal records check and no unmanageable debt. Suitability programs are commonly used when the consequences of relying on purely reactive security measures (described in detail later) are determined to be unacceptable. Suitability programs are found in disciplines wherein which we are unwilling to risk the consequences of relying solely on reactive measures, such as those dealing with children (early childhood care providers, all Head Start employees, teachers, school bus drivers), health (doctors, nurses, those who care for the elderly), homes (real estate brokers), money (accountants, brokers, financial advisors, bank employees, casino employees), and protection (law enforcement officers, security officers, investigators).

Suitability programs are carried out to protect the institution and the public from adverse consequences of inappropriate employee recruitment and retention. Suitability programs protect the institution by determining whether an employee is qualified to do a job. The benefits include reducing employee turnover, reducing risk of theft or embezzlement, preventing litigation over hiring practices, and reducing losses in productivity associated with recruitment and training of inappropriate personnel. Suitability programs protect the public by ensuring that only employees who are fit for their duties are employed in particularly risky endeavors. The programs are designed to encourage public trust by creating transparent rules and proactively mitigating risks.

In a laboratory biosecurity program, a suitability program can help to protect research from being compromised, either intentionally or inadvertently; to prevent proprietary or sensitive information from being released; to protect VBM from theft or misuse; and to protect laboratory personnel from threats, solicitation, or manipulation. These goals can be accomplished through prescreening credentials, criminal history, and references before employment and through ongoing monitoring of changes against an initial behavioral background. A suitability program does not ensure that an institution is fully protected and should be supplemented by personal, physical, information, and other security programs.

It is important to note that the Federal Bureau of Investigation's Security Risk Assessment (SRA), which is a requirement for to access to BSAT under the U.S. Select Agent Regulations (42 CFR 73, 9 CFR 121, 7 CFR 133), is not a comprehensive suitability program. The SRA is, in fact, a limited scope national database check that only covers a set of prohibitors defined in the Biological Weapons and Toxins Criminal Statute (18 U.S.C. 175), the U.S. Patriot Act (P.L. 107-56), and the Public Health Security and Bioterrorism Preparedness and Response Act of 2002 (P.L. 107-188 [13, 14, 15]). Negative information obtained in the SRA that does not meet the thresholds for the statutory "prohibitors" will not prevent an applicant from being granted access to BSAT.

In addition, privacy provisions of the select agent regulations prohibit the FBI from sharing personal information obtained in the SRA with the responsible official (RO). The Select Agent Program manages suitability programs for biological laboratories that conduct work with BSAT. This program serves to supplement the SRA by providing an institution with initial information upon initial screening that can then be monitored over time, and changes in status appropriately assessed.

As shown in Table 6.3 basic suitability programs can be built upon the foundations of the protective intelligence investigation, which was devised to identify and characterize a wide spectrum of violent threats. In the biomedical research, public health, and clinical communities the scope of the investigation can be refined to include items that further elucidate the applicant's bona fides and other potential indicators or stress or behaviors of concern.

Threat Assessment Programs

A threat assessment program expands upon the investigations carried out in a basic background check and is implemented when a direct, implied, or observed threat has occurred against the institution or any of its staff. Threat assessments can be used in the context of threat-driven physical security but can be even more effective in preventing acts against institutions when used as part of a threat assessment program. A threat assessment program includes three major functions:

1. Identification of a potential perpetrator
2. Assessment of the risks of violence posed by a given perpetrator at a given time
3. Management of both the subject and the risks that he, she or they present(s) to a given target

A threat assessment investigation is a key piece of a threat assessment program and is conducted to protect an institute or staff against a credible threat (e.g., an observed threat from an eco-terrorist group). Investigations may be conducted as part of pre-employment screening, enrollment in a biosecurity program to help protect individuals from coercion, or to establish a baseline of normal behavior when specified indicators or thresholds are breached in an initial suitability screening. Threat assessment investigations may also be conducted if an employee displays threatening behavior—for example, a laboratory worker who loses his/her temper and throws a glass flask containing bacterial culture or makes a direct threat by issuing a verbal or written statement such as, "Anyone can make a biological weapon. One of these days I am going to show the world how easy it is." In addition, threat assessment investigations may be conducted on potential victims that work in laboratories to determine vulnerabilities that require protection from a threat. For example, a stem-cell principal investigator that has received threatening mail at his or her residence because of the stem-cell work may require protection. A threat assessment investigation conducted on the stem-cell principal investigator can be used to determine key vulnerabilities that the adversary may exploit to do harm, and which the institution can subsequently develop countermeasures against to protect the worker. A threat assessment program can be incorporated into a biosecurity program so that the institution is adequately prepared to address potential threats. The following process can be used to build a threat assessment program:[7]

1. Define criteria that could lead to a person or group becoming a subject of a threat assessment investigation
2. Determine the areas within the law enforcement or security organization that will be responsible for receiving information about possible subjects and conducting threat assessment investigations
3. Notify those individuals and organizations that might come in contact with—or know of—potential subjects about the existence of a threat assessment program
4. Educate notified individuals and organizations about the criteria for bringing a concern about potential violence to the attention of investigators

External organizations (e.g., law enforcement assets or private investigation firms) can be called upon for help when a biosecurity program determines that a full-scale threat assessment investigation may be necessary.

There are several measures that will help in building a threat assessment program and in the process of identifying whether there is a potential threat (either internal or external) to the institution or its staff. First, it is important to establish a strong working relationship with your institution's security office. The security office should have established liaisons with local law enforcement and should be aware of any criminal activity or threats made against the institution. The security office should also be aware of and understand the value and risks associated with any VBM in order to evaluate potential external threats. Secondly, it is important to utilize existing safety and suitability programs as part of a threat assessment program. An increase in safety violations may indicate a security risk, which should be evaluated in consultation with the biosafety officer or responsible official. Human resources should be involved if there is a potential threat by or against an individual at your institute and can alert you of changes in the behavior of potential or suspected threats. This last facet of a successfully implemented threat assessment program highlights the need for a robust, proactive, real-time, self-reporting mechanism to bring changes in behavior to the attention of supervisors, security, and human resources professionals.[15]

Resources for Universities, Research Institutions, and Small Biotech Firms

The protective intelligence investigation was originally developed for small police departments and private security firms that provide executive protection services and as such was deliberately tailored to the needs of small organizations with limited threat assessment resources.[8] In these traditional security settings the approach has been very successful, resulting in subsequent adaptation to other security problems, including campus and school violence and the National Institutes of Health Intramural Personal Security Program.[9] In fact, many U.S. colleges and universities are already practicing a fairly advanced form of threat assessment through their campus violence prevention programs, created in the wake of the Virginia Tech massacre, Columbine High School shootings, Connecticut elementary school shooting, and other similar events. Most institutions, particularly large universities, have taken advantage of a diverse set of in-house subject matter experts (SMEs) to create a threat assessment "task force" composed of representatives from campus police/security, human resources, biosafety (responsible official if a regulated entity under the U.S. Select Agent Regulations), ombudsman, faculty, and in some cases occupational health psychiatrists and psychologists.[17]

In smaller institutions the principles of threat assessment presented here will serve as a very basic template for a security manager or responsible official to set up an initial threat assessment program. It is important to recognize the limitations of such a program and have a ready resource on-hand to intervene when an actual threat is identified or suspected. Local police, to some extent, provide this assistance, but a better resource is the local Federal Bureau of Investigation weapons of mass destruction (FBI WMD) coordinator. FBI WMD coordinators are special agents with WMD subject matter experstise, including a functional knowledge of biomedical research organizations, who serve at each of the 56 FBI field offices. Outside of law enforcement assets, a small research institution may call on partnering organizations such as larger regional centers of excellence at major research universities that may be able to share resources. Finally, there are private firms that provide contract security services. Security firms that specialize in executive or personal protection services can have extensive threat assessment capabilities. Not all private security firms are the same, and regulations governing their operations vary tremendously from state to state. A private security firm is, in many cases, the most efficient way to secure on-demand support that is tailored to the particular needs of the institution. However, not all private security firms are the same, and regulations governing their operations vary tremendously from state to state. Therefore, care should be taken when selecting and enlisting the support of a private security firm.

CONCLUSION

Having a firm knowledge of the potential threats to an institution and its people, and identifying strategies and tactics to mitigate those threats will greatly enhance the design and implementation of a biosecurity program and greatly improve an institution's capability to detect and address those threats. Like risk assessments, threat assessments require regular maintenance to ensure their relevance and viability. As a starting point, each institution should develop a custom baseline threat assessment and make the determination whether ongoing threat assessments are feasible. Regardless, understanding the basics of threat assessments will provide certain assurances for institutions conducting work on dangerous or VBM materials.

References

1. National Research Council Committee on Laboratory Security and Personnel Reliability Assurance Systems for Laboratories Conducting Research on Biological Select Agents and Toxins, Board on Life Science, Division of Earth and Life Sciences. Responsible Research with Biological Select Agents and Toxins. National Academies Press. Washington, DC (2009).
2. National Science Advisory Board for Biosecurity. Enhancing Personnel Reliability among Individuals with Access to Select Agents: A Report of the National Science Advisory Board for Biosecurity. National Institutes of Health, Office of Biotechnology Activities (2009). Available at: http://oba.od.nih.gov/biosecurity/meetings/200905T/NSABB%20Final%20Report%20on%20PR%205-29-09.pdf (accessed 2012).
3. National Science Advisory Board for Biosecurity Guidance for Enhancing Personnel Reliability and Strengthening the Culture of Responsibility. A Report of the National Science Advisory Board for Biosecurity. National Institutes of Health, Office of Biotechnology Activities (2011). Available at: http://oba.od.nih.gov/biosecurity/pdf/CRWG_Report_final.pdf (accessed 2012).
4. World Health Organization. WHO Biorisk Management: Laboratory biosecurity guidance (2006). Available at: http://www.who.int/csr/resources/publications/biosafety/WHO_CDS_EPR_2006_6.pdf.
5. Interagency Operation Security Support Staff National OPSEC Program. Glossary of terms. Available at: http://www.au.af.mil/au/awc/awcgate/documents/opsec_definitions.htm.

6. Bulling, D., and Scalora, M. Behavioral Science Guidelines for Assessing Insider Threats. University of Nebraska, Lincoln (2008). Available at: http://ppc.nebraska.edu/userfiles/file/Documents/projects/ThreatAssessment/BehavScienceGuidelinesforInsiderThreat.pdf (accessed 2012).
7. Fein, R.A., and Vossekuil, R. Protective Intelligence and Threat Assessment Investigations: A Guide for State and Local Law Enforcement Officials. National Institute of Justice (1998). Available at: http://www.secretservice.gov/ntac/PI_Guide.pdf (accessed 2012).
8. Fein, R.A., and Vossekuil, B. Assassination in the United States: an operational study of recent assassins, attacker and near-lethal approachers. J. Forensic. Sci. 44:321–333 (1999).
9. Skvorc, C., and Wilson, D.E. Biosecurity and Bioterrorism: Biodefense Strategy, Practice, and Science. 9: 23–29 (2011).
10. Fein, R.A., Vossekuil, R., and Holden, G.A. Threat Assessment: An Approach To Prevent Targeted Violence, National Institute of Justice (1995). Available at: http://www.secretservice.gov/ntac/ntac_threat.pdf (accessed 2012).
11. Federal Experts Security Advisory Panel. Recommendations Concerning the Select Agent Program (2011). Available at: http://www.phe.gov/Preparedness/legal/boards/fesap/Documents/fesap-recommendations-101102.pdf (accessed 2012).
12. Department of Health and Human Services Biosafety in Microbiological and Biomedical Laboratories, 5th edition. HHS Publication (CDC) 21-1112 (2009).
13. U.S. Department of Agriculture. Animal and Plant Health Inspection Service. APHIS Select Agents Regulations. Federal Register. 7 CFR Part 331 and 9 CFR Part 121 (2008).
14. Department of Health and Human Services. Centers for Disease Control. CDC Select Agents Regulations. Federal Register 42 CFR Part 73 (2008).
15. U.S. Patroit Act (U.S. H.R. 3162, Public Law 107-56), Title IX, Sec. 904-1011.
16. Virginia Tech Review Panel. Mass Shootings at Virginia Tech April 16, 2007: Report of the Review Panel Presented to Governor Kaine of the Commonwealth of Virgina (2007). Available at: http://www.governor.virginia.gov/tempcontent/techpanelreport.cfm (accessed 2012).
17. Berger, K.M., Wlinetz, C., McCarron, K., You, E., and So, W. Bridging Science and Security for Biological Research: A Dialogue Between Universities and the Federal Bureau of Investigation. Report of a Meeting February 21--22, 2012. American Association for the Advancement of Science (2012). Available at: http://cstsp.aaas.org/files/AAAS-AAU-APLU-FBI%20Final%20Report.pdf (accessed 2012).
18. National Research Council Committee on Review of the Scientific Approaches Used During the FBI's Investigation of the 2001 Bacillus antrhacis Mailings. Board on Life Sciences, Committee on Science, Technology and Law Policy and Global Affairs. Review of the Scientific Approaches Used During the FBI's Investigation of the 2001 Anthrax Letters. National Academies Press. Washington, DC (2011).
19. Expert Behavioral Analysis Panel Report of the Expert Behavioral Analysis Panel (unsealed redacted version) (2011). Available at: http://www.dcd.uscourts.gov/dcd/sites/dcd/files/unsealedDoc031011.pdf (accessed 2012).
20. Department of Justice (2011). Amerithrax Investigative Summary. Available at http://www.justice.gov/amerithrax/docs/amx-investigative-summary.pdf (accessed 2012).

21. Kolavic, S.A., Kimura, A., Simons, S.L., Slutsker, S., Barth, S., Haley, C.E. An outbreak of Shigella dysenteriae Type 2 among laboratory workers due to intentional food contamination. JAMA 278:369–398 (1997).
22. Kovaleski, S.F., and Robbins, L. Yale Killing Suspect Appears in Court, New York Times (2009). Available at: http://www.nytimes.com/2009/10/07/nyregion/07Yale.html (accessed 2012).
23. Carus, W.S. Working Paper, Bioterrorism and Biocrimes: The Illicit Use of BiologicalAgents Since 1900 (2001).
24. Bluestein, G., and Reeves, J. Associated Press (February 17, 2010). Available at: http://mdjonline.com/view/full_story/6375288/article-Professor-s-murky-past-raises-questions (accessed 2012).
25. Leonard, J., and Gottlieb, J. Meetings with South Africa Attache Under Scrutiny. Los Angeles Times (2000). Available at: http://articles.latimes.com/2000/jul/17/local/me-54309 (accessed 2012).
26. Thomas, J. California Doctor's Suicide Leaves Many Troubling Mysteries Unsolved. New York Times (2002). Available at: http://www.nytimes.com/2002/11/03/us/california-doctor-s-suicide-leaves-many-troubling-mysteries-unsolved.html?pagewanted=all&src=pm (accessed 2012).

PART III

BIOSECURITY IN VARIOUS SECTORS

CHAPTER 7

Biosecurity in Research Laboratories

Kavita M. Berger

INTRODUCTION

Ten years after the terrorist events of 2001, scientists face serious challenges in conducting research into biodefense and emerging infectious diseases, with ever-increasing regulation on research, facilities, and personnel. Concerns about the intentional misuse of information, research, biological materials, tools, and equipment to cause harm have prompted years of policy discussions, the development of educational tools and programs, and continuing U.S. policy on regulation of "dual-use research of concern." Fear that biological weapons could be used against the U.S. population have driven significant investment in biodefense research to help anticipate threats, conduct forensic analysis of microbial threats, and develop medical interventions to counter the health effects of potential biological weapons. The U.S. government has built high- and maximum-containment (BSL-3 and BSL-4, respectively) laboratory facilities and trained scientists to support this research. Conincident with this investment, concerns about theft of dangerous pathogens catalyzed the development of stringent policies, regulations, laws, and programs to limit access to those pathogens. Today, the U.S. finds itself in a paradoxical situation where significant financial and resources are devoted to biodefense research while bureaucratic resources focus on limiting access, theft, and misuse of knowledge and research. In some cases, the programs complement each other, but in others they are highly discordant. The consequence of this complex structure is systemic confusion among those involved and outside observers, as recently observed with two H5N1 papers. This chapter describes concerns about and challenges of biosecurity in research laboratories using the recent H5N1 controversy as a case study.

LABORATORY BIOSECURITY

In the United States, the term "laboratory biosecurity" refers to physical and personnel barriers that prevent deliberate release or theft of dangerous biological pathogens and toxins from laboratories. This differs from the concept of "biosafety," which refers to physical barriers and personnel training to prevent accidental release of dangerous pathogens into the environment or accidental exposure of researchers to harmful pathogens. The important distinction between these terms is the nature of the risk: i.e., is the release deliberate or accidental? The specific processes and procedures that research institutions must implement to address biosecurity and biosafety may overlap (i.e., using physical barriers to prevent unauthorized access to the facilities or ensuring that only authorized personnel have access to laboratories) or may be unique to one concept (i.e., closed-circuit televisions to monitor who is entering and exiting the laboratories—security—or directional air flow in the laboratory and hoods—safety. In recent years, the line between biosafety and biosecurity has blurred, and laboratory biosecurity was added to the fifth edition of Biosafety in Microbiological and Biomedical Laboratories (BMBL),[1] the Centers for Disease and Control and Prevention (CDC) guidance for laboratory biosafety. Consequently, officials who have responsibility for biosafety are also being tasked with addressing laboratory biosecurity in many research institutions.

Unlike the United States, much of the world does not distinguish laboratory biosecurity from biosafety. In fact, most languages do not have an equivalent term for "biosecurity." This has caused confusion and concern between scientists and security experts when discussing laboratory biosecurity with international partners. To minimize any confusion about the underlying concept of preventing deliberate and accidental release of dangerous pathogens from the laboratory, members of the biosafety and biosecurity communities are beginning to use the term "biorisk" and to develop processes and procedures for managing biorisks. An effort initiated through the European standardization organization has attempted to develop, harmonize, and standardize best practices for global biorisk management.[2] Despite international use of "biorisk" as the preferred term, the World Health Organization (WHO) has issued handbooks on laboratory biosecurity,[3] biosafety,[4] and guidance for responsible conduct,[5] which include principles of bioethics, biosafety, and biosecurity.

Laboratory biosecurity has become a critical part of research institutions. Although the overwhelming majority of research institutions (both public and private) do not work with BSATs, many of them encounter other security threats, including cyber security, animal rights extremism and violence, and campus violence.[6] Pharmaceutical and biotechnology companies have additional security concerns that involve protecting intellectual property and product secrets. Some offenders infiltrate themselves in the institutions before accomplishing their goals (i.e., the insider threat) while others attack institutions from outside (i.e., the external threat). Universities have created threat assessment teams to identify and/or alert the appropriate authorities in the university and within the neighboring community to threats that target the institution, staff, or students.[7] These threat assessment teams comprise the campus police or security force, neighboring community police, university officials, and other community members as appropriate. Universities are also employing various controls to prevent access of unauthorized individuals to research facilities.[8] Examples of these physical barriers include closed-circuit televisions, biometric controls on laboratory doors, key-card access, and security guards. Personnel with access to animal facilities and high-containment laboratories undergo initial and periodic training to ensure knowledge of the institution's policies and procedures for working safely in the facilities. This training helps prevent unauthorized access to facilities housing research animals.[9] Several universities have initiated background checks for prospective employees to prevent known offenders from gaining employment at research facilities.[10] In addition to these physical and personnel measures, many research institutions employ measures to keep track of their research materials and animals. These measures are far more stringent at pharmaceutical and biotechnology companies.

THE SELECT AGENT PROGRAM

Pathogens and toxins that pose significant risks to U.S. national security and public health are called "select agents" and are treated with greater security than other emerging infectious diseases. The select agent program began in the mid-1990s after two specific terrorist incidents: the 1995 acquisition of plague bacteria from the American Tissue Culture Collection by Larry Wayne Harris, a trained microbiologist and member of the Aryan Nation from Ohio; and the 1995 Oklahoma City bombing.[11] Following these incidents, Congress passed the Antiterrorism and Effective Death Penalty Act of 1996, which created the list of select biological agents and toxins and made transportation of these agents to unauthorized individuals illegal.[12] Shortly after the program was created, the United States was attacked by Al Qaeda and terrorized by letters containing anthrax spores. In the wake of the 2001 terrorist incidents, Congress passed the U.S. Patriot Act, which imposed restrictions on possession of select agents,[13] and the Public Health Security and Bioterrorism

Preparedness Act of 2002, which increased physical and personnel security requirements for select agents.[14] The select agent list expanded to include plant pathogens as well as human and animal pathogens and toxins, totaling 82 agents as of 2011.[15] The final select agent rules went into effect in 2005.[16]

The final rule of the select agent program enhanced security around pathogens and toxins on the list.[17] The CDC regulates select agents that infect people; the Animal and Plant Inspection Service (APHIS) regulates biological agents that infect animals and plants; and both agencies regulate pathogens that infect both people and animals (termed, "overlap agents"). The CDC and APHIS work together to develop program requirements and conduct joint inspections of facilities housing the overlap agents. The select agent program refers to the Biosafety in Microbiological and Biomedical Laboratories (BMBL) guidelines as standards for safety and security of personnel and facilities.[18] The CDC and APHIS also work with the Federal Bureau of Investigation (FBI) to approve facilities and individuals for access to select agents.[19] The FBI conducts security risk assessments (SRA), which consist of a series of database searches, including criminal, national security, and immigration checks.[20] All approved individuals and facilities are registered with the select agent program. Research institutions have particular staff, responsible official (RO) and alternate responsible official (ARO), dedicated to overseeing select agent laboratories and compliance with select agent program requirements.[21] Institutions also have security plans for their facilities and conduct periodic exercises to address gaps or vulnerabilities in security. These exercises often include key officials from the research institutions, select agent researchers, and campus security or police; increasingly, institutions are including members of the local police force and pubic health officials. Institutions implement performance-based physical security measures to prevent theft or otherwise unauthorized access to select agents.[22] Research institutions keep records of all select agents, including those received, created, used, transferred, or disposed of. Finally, institutions have procedures in place to handle reporting of illicit behavior. All of these requirements are included in the 2005 final select agent rules.

In 2007, news of an accidental exposure of a laboratory worker at Texas A&M University to *Brucella*, a select agent, prompted a series of policy discussions in the Executive Branch and Congress about the security of select agents.[23] In the Executive Branch, an interagency effort, led by representatives from the U.S. Departments of Health and Human Services and Agriculture, reviewed and provided recommendations on biosafety policies and procedures in high-containment laboratories.[24] These recommendations included improved coordination among federal agencies funding and overseeing research conducted in high-containment laboratories; development of a national strategy to ensure that all individuals working in or overseeing research in high-containment laboratories are competent in biosafety; collection and analysis of laboratory accidents and corrective actions to improve biosafety procedures; development of a national research agenda to improve laboratory management of biological hazards; and the sharing of public communication strategies.

In a parallel effort, other interagency groups were discussing laboratory biosecurity measures, including an ethical framework for select agent research, personnel security, and physical security.[25] This effort was initiated after the Congressional Commission on the Prevention of Weapons of Mass Destruction Proliferation and Terrorism (WMD Commission) published its 2008 report, World at Risk;[26] and Bruce Ivins, an anthrax researcher from the U.S. Army Medical Research Institute for Infectious Diseases (USAMRIID), was identified as the alleged perpetrator of the 2001 anthrax mailings.[27] In its report and subsequent Congressional hearing, the WMD Commission stated that a terrorist incident will occur in the United States by 2013 and that a biological attack was more likely to occur than a nuclear attack.[28] The WMD Commission advocated laboratory risk management to minimize the potential that biological research could be misused or dangerous pathogens could be stolen from laboratories. Coincident with the finalization of the commission's report, questions about the "insider threat" emerged as members of the security community learned about the FBI's case against Ivins, who had over 30 years of specialized experience working with anthrax.[29]

In response to these events, Congress introduced the WMD Prevention and Preparedness Act,[30] and President George W. Bush issued Executive Order 13486, Strengthening Laboratory Biosecurity in the United States.[31] Revisions have been proposed to the WMD Prevention and Preparendess Act several times since its initial introduction in 2009. The bill contains sections to enhance the select agent program by creating a top tier of high-priority threat agents and commensurate physical and personnel security measures, along with sections to improve preparedness and response capabilities, global biosecurity, and the intelligence workforce. The original bill also gave the Department of Homeland Security (DHS) responsibility for regulating security for the top tier agents.[32] Although the bill has been introduced several times during congressional sessions, it has yet to pass.[33] Executive Order 13486 called for a working group including members from all government agencies responsible for select agents or biosecurity to review physical, facility, and personnel security to address internal and external threats.[34] This working group provided recommendations on select agent regulations, personnel security, physical security, and transportation of select agents.[35] This report led to the issuance of Executive Order 13546, Optimizing the Security of Biological Select Agents and Toxins in the United States.[36] This order called for the creation of a tiered system of biological select agents with commensurate safety and security measures, coordination of requirements and inspections among all federal agencies with responsibilities for select agents, and the establishment of the Federal Experts Security Advisory Panel (FESAP). The FESAP was subsequently created to provide federal agencies with recommendations on how to implement the tiering system and enhance security.[37] After significant deliberation among panel members and with representatives from the research and security communities, the FESAP issued recommendations on the biological pathogens and toxins that should be in the top tier of select agents; the pathogens that should be removed from the current select agent list; personnel reliability, including improving the security risk assessments and providing guidance for initial assessments and periodic monitoring; and physical and cyber security to safeguard the select agents housed in high-containment laboratories.[38] The CDC and APHIS incorporated the FESAP recommendations in the biennal review of the select agent program.[39] The notice of the proposed rule making was released for public comment in the fall of 2011.

The Dual-Use Dilemma

Physical and personnel security and safety in laboratories are not the only biosecurity concerns associated with select agent research. For over a decade, members of the policymaking and security communities have been expressing concerns about the potential for legitimate research to be misused by individuals with malicious intent. Concerns that scientists could be wittingly or unwittingly coopted to help ill-intended individuals misuse biology have contributed to development of the concept, "dual-use dilemma." The term "dual-use research" was coined by the U.S. National Research Council in 2003 and refers to legitimate biological research, knowledge, or tools that could be misused by nefarious individuals to cause harm.[40] Subsequently, the U.S. government established an advisory group, the National Science Advisory Board for Biosecurity (NSABB), to determine the criteria, oversight policies, communication strategies, and education for "dual use research."[41] The NSABB released its initial recommendations in 2007,[42] and since then the U.S. government has been debating and developing an overarching policy to identify and handle "dual-use research."

In response to the recent H5N1 papers, the U.S. government has issued guidance for scientists and research sponsors to identify and modify experiments that may be considered "dual-use research of concern."[43] Guidance issued in April 2012 requires U.S. government agencies that fund research on high-priority select agents (15 are isted in the guidance) to review proposed research activities for their dual-use potential as defined by the NSABB's 2007 recommendations.[44] If dual-use research is identified in proposals, funding agencies must work with researchers to minimize risks by changing the experimental procedures, increasing biosafety and biosecurity, assessing the availability of medical interventions, reviewing research progress routinely, or using other strategies. This guidance offers a start toward addressing potential misuse of biological research, but gaps still remain in identifying, assessing, and communicating the risks of research.

Identifying research that qualifies as "dual-use research" may be challenging, because judgments often depend on familiarity with the experimental procedures in question, the pathogens involved, and national security risks. Internationally, scientists often confuse this concept with the dual benefits of the research project and knowledge.[45] Once the concept is defined as a security concern, many scientists in developing countries are offended by the thought that they are considered threats rather than positive contributors to science, health, and agriculture. Many American scientists have expressed the same concerns, but as their exposure to "dual-use" discussions increases, so does their acceptance of the concept.[46] Today, some scientists and research institutions in the United States have begun to review research for its potential to pose risks to national security; most of these individuals and entities struggle with assessing the level of security risk associated with research. The scientific community also faces challenges in communicating the risks of research activities and strategies used to minimize potential security risks to the sponsors, security community, and general public.

Biosecurity Outreach

For several years, the U.S. government has engaged in efforts to reach out to the scientific community to raise awareness about biosecurity and ensure that authoritative information is available. The NSABB has participated in conferences to discuss the recommendations and has invited members of the scientific community to provide their experiences with "dual-use research," responsible conduct, synthetic genomics, and personnel reliability. The White House Office of Science and Technology Policy and the Office of the Director of National Intelligence led an effort to develop a conference display and website that provide authoritative information about all relevant policies and programs related to biosecurity.[47] The White House has a website that defines and describes all biosecurity terms and programs.[48] The FBI WMD Directorate's biosecurity outreach initiative helps the scientific community mitigate challenging security problems that arise at research institutions.[49] The FBI's WMD coordinators in the 56 field offices and headquarters actively engage scientists and institutional officials in the amateur and professional (both public and private sector) communities. They have begun to build trust and partnerships with research institutions by helping to counter threats posed by violent animal rights extremists, campus violence, and cyber security. The FBI WMD coordinators have reached out to the amateur community by promoting responsible behavior and appealing to their interest in harmlessly "playing" with biology.[50] Interactions between the FBI and the amateur and professional scientific communities are steadily increasing, and the FBI is viewed as a helpful partner in ensuring that legitimate, beneficial science continues.

The H5N1 Papers

In 2011, after years of reaching out to scientists, research administrators, and journal editors, the NSABB was asked to assess the security risks of two recently submitted manuscripts.[51] Two independent research groups—one in the United States and the other in the Netherlands—submitted manuscripts that described strains of influenza that could easily spread between ferrets in the laboratory to two premier scientific journals, *Nature* and *Science*. These studies were conducted as part of the U.S. biodefense and emerging infections research portfolio and sought to identify the mutations in the H5 hemagglutinin protein that could enable naturally circulating H5N1 avian influenza viruses to easily spread among humans.[52] The justification for such studies was to provide public health officials with specific genetic or protein markers that could be used to identify H5N1 influenza viruses posing significant public health risks and to develop vaccines and drugs that would be effective against H5 viruses containing these mutations. The American group recognized the risks involved in conducting such research and chose to minimize those risks by using a 2009 H1N1 viral vector containing the H5 hemagglutinin protein and carrying out the research under appropriate biosafety and biosecurity conditions.[53] The 2009 H1N1 influenza virus was viewed as "less risky" because effective vaccines and drugs were available to prevent infection and treat disease. In contrast, the Dutch group used an H5N1 influenza virus in its studies.[54] Since the grants were funded, the scientists had discussed the research and results at scientific conferences and other similar venues prior to submitting the manuscripts to the journals. It was not until the papers were submitted in the fall of 2011 that they were flagged as posing potential security risks.

After being flagged by *Science* and *Nature*, the papers were reviewed by the NSABB for their potential to convey information that could enable known and would-be terrorist entities to develop an engineered biological weapon with H5N1 influenza virus. NSABB members agreed about the fate of the American paper but disagreed about the Dutch paper.[55] The NSABB recommended that the papers be published only after key information about the mutations and the methods used to create them was removed.[56] They also recommended that the redacted information should be shared with individuals who demonstrate a legitimate "need to know" about this information. This prompted the Dutch government to impose export controls on the information contained in the Dutch group's manuscript.[57] In February 2012, WHO convened an ad hoc group to evaluate the risks and benefits of the research.[58] It concluded that the papers provided a great benefit to global health and pandemic influenza preparedness efforts and recommended that they be published in full at a later date rather than immediately in partial form.[59] WHO supported the moratorium on the research, already initiated by the researchers themselves, until a clear understanding of how to proceed responsibly was determined.

Based on the NSABB and WHO recommendations, the authors revised their manuscripts. In March 2012, the NSABB reviewed the revised manuscripts and unanimously concluded that the papers no longer posed significant risks to national security and should be published.[60] However, the Dutch government did not lift its export control requirements until late April 2012, when it granted an export license that would allow *Science* to receive and publish the information.[61] The American group's paper, along with an accompanying editorial, was published in *Nature* in early May 2012.[62] The editorial supported full publication of any research that benefits society and public health.[63] Only time will tell if the information conveyed in the manuscripts contribute to global efforts to identify, monitor, and mitigate highly pathogenic avian influenza or lead to the development of engineered biological weapons.

Two major lessons emerged from the H5N1 papers. For decades, scientists have been grappling with evaluating and balancing risks and benefits of research. Until the last decade, most risks were associated with ethical and environmental concerns. Often data is available to help scientists assess ethical and environment risks and identify possible ethical considerations that might be relevant to the research activities in question. During the past decade, the scientific community has been asked to assess the potential that their research may pose risks to national security. Unlike other risks, no data or literature is publicly available to help scientists determine the likelihood and magnitude of security risk or benefit. Assessment of risks and benefits are highly subjective and influenced by one's laboratory experiences and familiarity with biosecurity and/human, animal, or plant health issues. Addressing the need for more clarity on how to identify risks and benefits and conduct risk-benefit evaluations in the absence of concrete data will be necessary to reduce potential national security risks associated with biological research prior to the publication stage.

In 2007, the NSABB recommended that scientists and institutions develop communication strategies to deal with publication of research that could pose security risks.[64] Although these tactics have been used in the past, they were insufficient in the recent H5N1 manuscript situation. Alarmist and accusatory rhetoric was published in the popular press by members of the media, whereas more thoughtful articles were published in scientific journals. At the 2012 AAAS Science and Technology Forum, David Brown from the *Washington Post* said that one of the major challenges the mass media faced at the beginning of the situation was that the scientists were not speaking with the media, which contributed to some of the misinformation in the popular press. For example, the press claimed that the modified H5 viruses were not only transmissible between ferrets but that they were also lethal, as was later revealed; the modified viruses were not lethal. Soon after the NSABB issued the first recommendation, the Dutch scientist became outspoken about the experiments and information in his manuscripts, the NSABB's review process and recommendations, and the Dutch government's issuance of export controls. This type of communication can breed misinformation about the processes associated with the immediate situation and responsibility of researchers to national security and public health.

CONCLUSION

Biosecurity in research laboratories encompasses a broad range of issues, including physical and personnel security, misuse of biological knowledge and research, responsible conduct, and cyber security. Many of these issues have been the subject of policy debates and programmatic activities in the United States and internationally. As these discussions evolve, so do the cadre of stakeholders engaged in the debates. Unlike a decade ago, the scientific, public health, administrative, and biosafety communities are not only informed about biosecurity issues but are actively included in timely policy debates. This inclusion of diverse stakeholders has enriched the security and policymaking communities' understanding of laboratory biosecurity, biosafety, and "dual-use research." Continuing in this direction will ensure that appropriate, evidence-based security measures are employed at research institutions to minimize both domestic and national security threats and maximize limited financial, administrative, and bureaucratic resources.

References

1. Access at: http://www.cdc.gov/biosafety/publications/bmbl5/.
2. European Committee for Standardization. CEN Workshop Agreement: Laboratory Biorisk Management. CWA 15793 (2008). Access at: ftp://ftp.cenorm.be/CEN/Sectors/TCandWorkshops/Workshops/CWA15793_September2011.pdf.
3. World Health Organization. Biorisk Management: Laboratory Biosecurity Guidance (2006). Access at: http://www.who.int/csr/resources/publications/biosafety/WHO_CDS_EPR_2006_6.pdf.
4. World Health Organization. Laboratory Biosafety Manual, 3rd edition (2004). Access at: http://www.who.int/csr/resources/publications/biosafety/Biosafety7.pdf.
5. World Health Organization. Responsible Life Sciences Research for Global Health Security: A Guidance Document (2010). Access at: http://whqlibdoc.who.int/hq/2010/WHO_HSE_GAR_BDP_2010.2_eng.pdf.
6. American Association for the Advancement of Science, Association of American Universities, and Association of Public- and Land-grant Universities. Competing Responsibilities?: Addressing the Security Risks of Biological Research in Academia (2010). Access at: http://cstsp.aaas.org/files/Competing%20Responsibilities%20AAAS%20AAU%20APLU%20report%20final.pdf; Massachusetts Society for Medical Research and Ohio Scientific Education and Research Association. Security. Conference, Columbus, OH (2011) . Access at: http://www.msmr.org/workshops.html; Massachusetts Society for Medical Research. Harmonization of the IACUCs, IBCs and IRBs. Tempe, AZ (2012). Access at: http://www.msmr.org/workshops.html.
7. American Association for the Advancement of Science, Association of American Universities, and Association of Public- and Land-grant Universities. Competing Responsibilities?: Addressing the Security Risks of Biological Research in Academia (2010). Access at: http://cstsp.aaas.org/files/Competing%20Responsibilities%20AAAS%20AAU%20APLU%20report%20final.pdf; American Association for the Advancement of Science, Association of American Universities, Association of Public- and Land-grant Universities, and the Federal Bureau of investigation. Bridging Science and Security for Biological Research: A Dialogue between University and the Federal Bureau of Investigation (2012). Access at: http://cstsp.aaas.org/files/AAAS-AAU-APLU-FBI%20Final%20Report.pdf.
8. Ibid.
9. American Association for the Advancement of Science, Association of American Universities, and Association of Public- and Land-grant Universities. Competing Responsibilities?: Addressing the Security Risks of Biological Research in Academia (2010). Access at: http://cstsp.aaas.org/files/Competing%20Responsibilities%20AAAS%20AAU%20APLU%20report%20final.pdf.
10. American Association for the Advancement of Science, Association of American Universities, Association of Public- and Land-grant Universities, and the Federal Bureau

of investigation. Bridging Science and Security for Biological Research: A Dialogue between University and the Federal Bureau of Investigation (2012). Access at: http://cstsp.aaas.org/files/AAAS-AAU-APLU-FBI%20Final%20Report.pdf.

11. U.S. Government. Report of the Working Group on Strengthening the Biosecurity of the United States (2009). Access at: http://www.phe.gov/Preparedness/legal/boards/biosecurity/Documents/biosecreportfinal102309.pdf.
12. Antiterrorism and Effective Death Penalty Act of 1996. P.L. 104-132. Title V, Subtitle B.
13. Uniting and Strengthening America by Providing Appropriate Tools Required to Intercept and Obstruct Terrorism (U.S. Patriot) Act of 2011. P.L. 107-56. Title VIII, Sec 817.
14. Public Health Security and Bioterrorism Preparedness Act of 2002. P.L. 107-188. Title III.
15. The select agent list is available at: http://www.selectagents.gov/select%20agents%20and%20Toxins%20list.html.
16. 42 CFR Parts 72 and 73, Office of the Inspector General, 42 CFR Part 1003. Possession, Use, Transfer of Select Agents and Toxins (2005). Access at: http://www.selectagents.gov/resources/42_cfr_73_final_rule.pdf.
17. Ibid.
18. Blaine, J. Facility Inspections Under the CDC Select Agent Rule. Applied Biosafety, 9(2), 84–87 (2004); Besser, R.E. Testimony: Oversight of Select Agents by the Centers for Disease Control and Prevention (2007). Access at: http://www.hhs.gov/asl/testify/2007/10/t20071004c.html.
19. 42 CFR Parts 72 and 73, Office of the Inspector General, 42 CFR Part 1003. Possession, Use, Transfer of Select Agents and Toxins (2005). Access at: http://www.select agents.gov/resources/42_cfr_73_final_rule.pdf.
20. Ibid.
21. Ibid.
22. American Association for the Advancement of Science, Association of American Universities, Association of Public- and Land-grant Universities, and the Federal Bureau of investigation. Bridging Science and Security for Biological Research: A Dialogue between University and the Federal Bureau of Investigation (2012). Access at: http://cstsp.aaas.org/files/AAAS-AAU-APLU-FBI%20Final%20Report.pdf.
23 U.S. Government. Report of the Trans-Federal Task Force on Optimizing Biosafety and Biocontainment Oversight (2009). Access at: http://www.ars.usda.gov/is/br/bbotaskforce/biosafety-FINAL-REPORT-092009.pdf; U.S. House of Representatives Committee on Energy and Commerce. Germs, Viruses, and Secrets: The Silent Proliferation of Bio-Laboratories in the United States (2007). Access at: http://www.gpo.gov/fdsys/pkg/CHRG-110hhrg44948/html/CHRG-110hhrg44948.htm.
24. U.S. Government. Report of the Trans-Federal Task Force on Optimizing Biosafety and Biocontainment Oversight (2009). Access at: http://www.ars.usda.gov/is/br/bbotaskforce/biosafety-FINAL-REPORT-092009.pdf.
25. President George. W. Bush. Executive Order 3486: Strengthening Laboratory Biosecurity in the United States (2009). Access at: http://www.gpo.gov/fdsys/pkg/FR-2009-01-14/pdf/E9-818.pdf; President Barack Obama. Executive Order 13546: Optimizing the Security of Biological Select Agents and Toxins in the United States (2010). Access at: http://www.gpo.gov/fdsys/pkg/FR-2010-07-08/pdf/2010-16864.pdf.

26. Commission on the Prevention of Weapons of Mass Destruction Proliferation and Terrorism. World at Risk (2008). Access at: http://www.absa.org/leg/WorldAtRisk.pdf; Senate Committee on Homeland Security and Governmental Affairs. World at Risk: A Report from the Commission on the Prevention of Weapons of Mass Destruction Proliferation and Terrorism (2008). Access at: http://www.hsgac.senate.gov/hearings/world-at-risk-a-report-from-the-commission-on-the-prevention-of-weapons-of-mass-destruction-proliferation-and-terrorism.
27. New York Times. Bruce E. Ivins (2008). Access at: http://topics.nytimes.com/top/reference/timestopics/people/i/bruce_e_ivins/index.html.
28. Commission on the Prevention of Weapons of Mass Destruction Proliferation and Terrorism. World at Risk (2008). Access at: http://www.absa.org/leg/WorldAtRisk.pdf; Senate Committee on Homeland Security and Governmental Affairs. World at Risk: A Report from the Commission on the Prevention of Weapons of Mass Destruction Proliferation and Terrorism (2008). Access at: http://www.hsgac.senate.gov/hearings/world-at-risk-a-report-from-the-commission-on-the-prevention-of-weapons-of-mass-destruction-proliferation-and-terrorism.
29. Imperiale, M.J., and Casadevall, A. Bioterrorism: Lessons Learned Since the Anthrax Mailings. mBio, 2(6), e00232-11 (2011).
30. WMD Prevention and Preparedness Act 2009. S.1649; WMD Prevention and Preparedness Act. H.R.2356 2011.
31. President George. W. Bush. Executive Order 3486: Strengthening Laboratory Biosecurity in the United States (2009). Access at: http://www.gpo.gov/fdsys/pkg/FR-2009-01-14/pdf/E9-818.pdf.
32. WMD Prevention and Preparedness Act. S.1649 (2009).
33. WMD Prevention and Preparedness Act. H.R.2356 (2011). At the time of the writing of this chapter, the bill will be discussed on the House of Representatives Floor. Access at: http://www.cidrap.umn.edu/cidrap/content/bt/bioprep/news/may1012biodefense.html.
34. President George. W. Bush. Executive Order 3486: Strengthening Laboratory Biosecurity in the United States (2009). Access at: http://www.gpo.gov/fdsys/pkg/FR-2009-01-14/pdf/E9-818.pdf.
35. U.S. Government. Report of the Working Group on Strengthening the Biosecurity of the United States (2009). Access at: http://www.phe.gov/Preparedness/legal/boards/biosecurity/Documents/biosecreportfinal102309.pdf.
36. President Barack Obama. Executive Order 13546: Optimizing the Security of Biological Select Agents and Toxins in the United States (2010). Access at: http://www.gpo.gov/fdsys/pkg/FR-2010-07-08/pdf/2010-16864.pdf.
37. Federal Experts Security Advisory Committee. Access at: http://www.phe.gov/Preparedness/legal/boards/fesap/Pages/default.aspx.
38. U.S. Government. Federal Experts Security Advisory Panel: Recommendations Concerning the Select Agent Program (2011). Access at: http://www.phe.gov/Preparedness/legal/boards/fesap/Documents/fesap-recommendations-101102.pdf.
39. 42 CFR Part 73. Possession, Use, and Transfer of Select Agents and Toxins; Biennial Review; Proposed Rule (2011). Access at: http://www.selectagents.gov/resources/HHS%20Federal%20Register.pdf; 7 CFR Part 331 and 9 CFR Part 121. Agricultural Bioterrorism Protection Act of 2002; Biennial Review and Republication of the Select Agent and Toxin List; Amendments to the Select Agent and Toxin Regulations; Proposed Rule (2011). Access at: http://www.selectagents.gov/resources/USDA%20Federal%20Register.pdf.

40. National Research Council. Biotechnology Research in an Age of Terrorism. National Academies Press. Washington, DC (2004).
41. National Science Advisory Board for Biosecurity. Access at: http://oba.od.nih.gov/biosecurity/about_nsabb.html.
42. National Science Advisory Board for Biosecurity. Proposed Framework for the Oversight of Dual Use Life Sciences Research: Strategies for Minimizing the Potential Misuse of Research Immunization (2007). Access at: http://oba.od.nih.gov/biosecurity/pdf/Framework%20for%20transmittal%200807_Sept07.pdf.
43. U.S. Government. U.S. Government Policy for Oversight of Life Sciences Dual Use Research of Concern (2012). Access at: http://oba.od.nih.gov/oba/biosecurity/PDF/United_States_Government_Policy_for_Oversight_of_DURC_FINAL_version_032812.pdf.
44. Ibid.
45. American Association for the Advancement of Science and Kuwait Institute for Scientific Research. International Engagement: Responsible Bioscience for a Safe and Secure Society. Workshop Two (2011). Access at: http://cstsp.aaas.org/InternationalMeeting/Kuwait/InternationalEngagementWS2.pdf; American Association for the Advancement of Science and Institute Pasteur de Tunis. Infectious Diseases and International Engagement: Responsible Bioscience for a Safe and Secure Society. Workshop Three (2012). Access at: http://cstsp.aaas.org/InternationalMeeting/Tunisia/Tunisreportfinal.pdf.
46. American Association for the Advancement of Science, Association of American Universities, Association of Public- and Land-grant Universities, and the Federal Bureau of investigation. Bridging Science and Security for Biological Research: A Dialogue between University and the Federal Bureau of Investigation (2012). Access at: http://cstsp.aaas.org/files/AAAS-AAU-APLU-FBI%20Final%20Report.pdf.
47. Science, Safety, and Security: Finding Balance Together. Access at: http://phe.gov/s3/Pages/default.aspx.
48. White House Office of Science and Technology Policy. Biosecurity FAQ. Access at: http://www.whitehouse.gov/administration/eop/ostp/nstc/biosecurity.
49. FBI Academic Biosecurity Workshops. Access at: http://academicbiosecurityworkshop.org/.
50. Lempinen, E. FBI, AAAS Collaborate on Ambitious Outreach to Biotech Researchers and DIY Biologists (2011).
51 National Institutes of Health. Press Statement on the NSABB Review of H5N1 Research (2011). Access at: http://www.nih.gov/news/health/dec2011/od-20.htm; Berns, K.I., Casadevall, A., Cohen, M.L., Ehrlich, S.A., Enquist, L.W., Fitch, J.P., Franz, D.R., Fraser-Liggett, C.M., Grant, C.M., Imperiale, M.J., Kanabrocki, J., Keim, P.S., Lemon, S.M., Levy, S.B., Lumpkin, J.R., Miller, J.F., Murch, R., Nance, M.E., Osterholm, M.T., Relman, D.A., Roth, J.A., Vidaver, A.K. Adaptations of avian flu virus are a cause for concern. Science, 335(6069), 660-1 (2012).
52. NSABB Members. Personal Communication (May 2012).
53. Kawaoka, Y. Transmission of an Influenza Virus Possessing and H5 Hemagglutinin via Respiratory Droplets in Ferrets. Royal Society meeting on H5N1 Research: Biosafety, Biosecurity, and Bioethics (2012). Access at: http://www.voiceprompt.co.uk/royalsociety/030412/.
54 Fouchier, R. Aerosol Transmission of Influenza A/H5N1 Virus in Ferrets. Royal Society meeting on H5N1 Research: Biosafety, Biosecurity, and Bioethics (2012). Access at: http://www.voiceprompt.co.uk/royalsociety/030412/.

55. NSABB members were evenly divided on whether to publish the Dutch paper.
56. National Institutes of Health. Press Statement on the NSABB Review of H5N1 Research (2011). Access at: http://www.nih.gov/news/health/dec2011/od-20.htm; Berns, K.I., Casadevall, A., Cohen, M.L., Ehrlich, S.A., Enquist, L.W., Fitch, J.P., Franz, D.R., Fraser-Liggett, C.M., Grant, C.M., Imperiale, M.J., Kanabrocki, J., Keim PS, Lemon SM, Levy SB, Lumpkin JR, Miller JF, Murch R, Nance ME, Osterholm MT, Relman, D.A., Roth, J.A., Vidaver, A.K. Adaptations of avian flu virus are a cause for concern. Science, 335(6069), 660-1 (2012).
57. Greenfieldboyce, N. Bird Flu Studies Mired in Export Control Law Limbo. NPR (2012). Access at: http://www.npr.org/blogs/health/2012/04/10/150311034/bird-flu-studies-mired-in-export-control-law-limbo.
58. Cohen, J. WHO Group: H5N1 Papers Should Be Published In Full. Science, 35(6071), 899-900 (2012).
59. World Health Organization. Public Health, Influenza Experts Agree H5N1 Research Critical, but Extend Delay (2012). Access at: http://www.who.int/mediacentre/news/releases/2012/h5n1_research_20120217/en/index.html.
60. National Science Advisory Board for Biosecurity. Statement of the NSABB. March 29–30, 2012, Meeting of the National Science Advisory Board for Biosecurity to Review Revised Manuscripts on Transmissibility of A/H5N1 Influenza Virus (2012). Access at: http://oba.od.nih.gov/oba/biosecurity/PDF/NSABB_Statement_March_2012_Meeting.pdf.
61. Enserink, M. Dutch Government OK's Publication of H5N1 Paper. ScienceInsider (2012). Access at: http://news.sciencemag.org/scienceinsider/2012/04/dutch-government-oks-publication.html.
62. Enserink, M., and Cohen, J. One H5N1 Paper Finally Goes to Press; Second Greenlighted. Science, 336(6081), 529-530 (2012); Imai, M., Watanabe, T., Das, S.C., Ozawa, M., Shinya, K., Zhong, G., Hanson, A., Katsura, H., Watanabe, S., Li, C., Kawakami, E., Yamada, S., Kiso, M., Suzuki, Y., Mahar, E.A., Neumann, G., Kawaoka, Y. Experimental Adaptation of an Influenza H5 HA Confers Respiratory Droplet Transmission to a Reassortant H5 HA/H1N1 Virus in Ferrets. Nature, doi:10.1038/nature10831 (2012). Access at: http://www.nature.com/nature/journal/vaop/ncurrent/pdf/nature10831.pdf.
63. Publishing Risky Research. Nature, 485(8). Access at: http://www.nature.com/nature/journal/v485/n7396/full/485005a.html.
64. National Science Advisory Board for Biosecurity. Proposed Framework for the Oversight of Dual Use Life Sciences Research: Strategies for Minimizing the Potential Misuse of Research Immunizatio(2007) . Access at: http://oba.od.nih.gov/biosecurity/pdf/Framework%20for%20transmittal%200807_Sept07.pdf.

CHAPTER 8

Biosecurity in the Food and Agricultural Industries

Henry S. Parker

INTRODUCTION

A major disruption to the U.S. food supply would have devastating consequences. Yet, only a decade ago, the nation's food and agriculture sector was not even designated as a critical national infrastructure.[1] Americans took for granted their abundant, safe, and affordable food resources.

September 11, 2001 confronted Americans with the harsh reality that the country's vital interests—including those on our own soil—were at risk from attacks from a small cadre of committed terrorists. Americans learned that terrorists were even targeting the nation's food supply. In 2002 U.S. and allied troops uncovered, in remote caves in Afghanistan, documents suggesting Al Qaeda was contemplating attacks against U.S. crops and livestock.[2]

In the years since the catastrophic events of 9/11, the U.S. has made substantial progress in its ability to protect its citizens and infrastructure against attacks from terrorists. Yet the nation's food supply remains at substantial risk from both intentional and unintentional disruption.

THE U.S. FOOD AND AGRICULTURE INFRASTRUCTURE

The U.S. food and agriculture infrastructure is extensive, complex, highly successful, and critically important for national security. Yet, those very characteristics mark the U.S. food supply as a tempting target. Effective protection of our agricultural resources and of the citizens that rely on them requires a comprehensive understanding of the nation's food production, processing, distribution, and sales systems.

Extent and Status of U.S. Agriculture

Food Security and Environment

The U.S. has the most extensive and successful agricultural enterprise on earth, and it is vitally important not only to the health and well-being of Americans but to the natural environment. It is the foundation of our food security. Food security has been defined as "physical and economic access to sufficient, safe, and nutritious food that meets the dietary needs and food preferences for an active and healthy life."[3] Access, by implication, means both affordability and ready availability.

Economic Impact

The U.S. agricultural enterprise has a substantial economic impact. In 2011, farmers' estimated cash receipts from livestock and crops were $363 billion—$166 billion from livestock and $197 billion from crops.[4] Commodities other than livestock and crops constitute a major component of U.S. agricultural production. These include products from forestry activities; nursery, landscape, and floral enterprises; more diverse agriculture-related businesses such as aquaculture, beekeeping, and maple sugaring; and plants grown for energy (approximately one-third of the nation's corn is now grown for ethanol production).[5]

These upstream and downstream food- and agriculture-related activities contribute far more to the economy than the on-farm cash receipts. Direct sales of commodities account for less than one percent of the nation's gross domestic product (GDP); however, if the total economic impact of the entire spectrum of agricultural and food-related industries is considered, the total contribution to GDP is over four percent.[4] Also, the total spectrum of food, fiber, and related industries provides employment for nearly 24 million Americans—20 percent of U.S. civilian employment.[4]

Exports are a large component of the economic impact of U.S. food and agriculture. First, agricultural exports generate a large economic multiplier. The economic multiplier associated with an agricultural commodity represents the total economic activity deriving from that commodity and includes the commodity's farm gate value plus the value emanating from post-harvest activities relating to that commodity, including transportation, marketing, and processing.[1] With regard to the export of U.S. agricultural commodities, every dollar of the commodity value yields an additional $1.34 in economic activity, and every one billion dollars in agricultural exports is supported by 7,800 U.S. jobs.[6] Second, U.S. agriculture is consistently a positive contributor to the nation's trade balance. The U.S. regularly exports a substantially greater value of agricultural products than it imports. For example, in 2011, the value of exports was over $137 billion, while the value of imports was only $95 billion—a trade surplus of $42 billion.[6] This surplus comes into sharp relief when contrasted with the trade *deficit* related to nonagricultural products: nearly $900 billion in 2011.[7]

Principal Characteristics of the U.S. Agricultural Infrastructure

The U.S. agricultural infrastructure is characterized by its complexity, diversity, innovative nature, high degree of concentration, vertical integration, and resilience. The complexity of U.S. agriculture is related to the number, sophistication, and interrelationships of the different systems, facilities, and activities that support the nation's agricultural industry.

The diversity of U.S. agriculture is manifested primarily in the large variety of the commodities produced and of the systems involved in the production, harvesting, processing, transportation, distribution, and sale of the commodities. However, U.S. agriculture tends to be highly concentrated, both economically and geographically, posing a potential security dilemma. The concentrated nature of U.S. agriculture also extends to densely packed populations of crops in the field, of livestock in feedlots, and of poultry in broiler houses, as well as to assemblages of postharvest food items. While such concentrations may enhance production and distribution efficiency, they can also accelerate the spread of disease or contamination.

The U.S. agriculture industry is also characterized by widespread vertical integration, meaning that a single company often controls much of the production, processing, distribution, and even sales of a commodity.[1]

Finally, the U.S. agricultural infrastructure is fundamentally resilient, as consumers can usually overcome supply shortfalls of specific commodities due to poor weather or other disruptions. Also, consumers can usually substitute cheaper or more abundant alternatives (chicken for beef, for example). Resilience is an important characteristic of a successful agricultural enterprise. This concept will be discussed in more detail later in the chapter.

The system's entire continuum, from producer to consumer, is highly productive, efficient, affordable, safe, innovative, and resilient. Yet, the industry faces significant challenges. National and global population growth coupled with increasingly scarce amounts of arable land, finite availability of suitable water resources, and the largely unpredictable consequences of global climate change threaten future food security. The growing competition among food production, bioenergy production, and other uses of farmland is already affecting supplies, prices, and outlets for some commodities. Rising energy costs and an increasingly complex regulatory environment are challenging the agricultural industry; inevitably raising product costs. Perhaps the greatest concern is the expanding arsenal of pathogens and toxic agents threatening the entire U.S. agricultural infrastructure.

FOOD DEFENSE

Protection of the U.S. food and agricultural system requires a comprehensive set of defensive measures to protect its fundamental elements from attack. Collectively, these measures comprise a commitment to *food defense*. To understand the concept of food defense; to distinguish food defense from other concepts related to food abundance, accessibility, quality, and safety; and to consider food defense in the context of terrorism first require a brief presentation of terminology and definitions. That will be followed by background discussion of biological agents and bioterrorism, a short history of bioterrorism with a specific focus on agroterrorism, and a consideration of the current and future threat of agroterrorism in the U.S.

Food defense has been described as "the collective term used by the FDA, USDA, DHS, etc., to encompass activities associated with protecting the nation's food supply from deliberate or intentional acts of contamination or tampering."[8] In this sense, food defense employs components of both biosecurity and biosafety, as defined and discussed in earlier chapters of this book. Food defense may be distinguished from food security ("physical and economic access to sufficient, safe, and nutritious food that meets the dietary needs and food preferences for an active and healthy life") and food safety ("the protection of food products from unintentional contamination by agents reasonably likely to occur in the food supply").[9] In fact, some have argued that the concept of food defense should be expanded to include both deliberate and unintentional threats and events. For example, William Ramsey, Chairman of the Food Defense Committee of the Food Products Association, stated, "Food defense ... is a compilation of all the efforts needed to protect the products—and therefore the business—from anything or anyone that could harm them."[10] This chapter will focus on the narrower concept of food defense as protection against deliberate attacks, even while recognizing that effective food defense will help protect the food supply from unintentional threats and events as well.

Agroterrorism is the generally accepted term used to describe terrorism directed against the food and agriculture infrastructure. While terrorists could physically disrupt the infrastructure (i.e., by destruction of facilities or transportation networks) or employ chemical or radiological contaminants, the greatest threat is *bioterrorism*—involving the use of biological agents against food and agriculture targets.

Biological Agents, Bioterrorism, and Agriculture

Biological Agents as Anti-Agriculture Weapons

Biological agents are often highly contagious, with effects ranging from debilitation to lethality. The foot and mouth disease virus, for example, is one of the most contagious pathogens. While it does not kill exposed cloven-hoofed animals, the debilitating effects of the disease make the livestock unmarketable.

Biological agents are often low-cost, high-consequence weapons and as such advantageous to terrorists. In 1969, a United Nations Chemical-Biological Expert Panel asserted, "for a large-scale operation against a civilian population, casualties might cost about $2,000 per square kilometer with conventional weapons, $800 with nuclear weapons, $600 with nerve-gas weapons, and $1 with biological weapons."[11,12] Biological agents used against agriculture have an additional "benefit" that, if used against crops and farm animals, they would likely be less risky to human handlers (unless a zoonotic agent is employed) and may require less technical knowledge to procure and deploy. (For example, a terrorist could, with little difficulty or risk, take a swab of foot and mouth disease virus from an infected animal overseas, smuggle the swab into the U.S., and covertly apply it to a susceptible farm animal here.)

Still, biological agents, including those posing risks on agriculture, are not ideal weapons. To attain the largest dispersal radius, aerosolizing and atmospheric release of the agent would be most effective. However, controlling the dispersal path could be significantly complicated by weather conditions, and microbes would potentially lose virulence or be rendered ineffective due to desiccation or exposure to ultraviolet radiation.[1] In addition, it is difficult to viably store biological agents, to preserve their quality, and to maintain them in a delivery state. Finally, using water or food as a vector for the agent can be ineffective. Pathogen or toxin concentrations would be significantly diluted in water, and cooking food would, in most cases, inactivate the agent.

Bioterrorists seeking to contaminate the food supply have a host of potential agents at their disposal. These include various viruses, bacteria, and rickettsia; prions; fungi; biotoxins; protozoans; and macro-organisms such as insects and weeds. Genetically modified organisms that combine the insidious effects of two or more agents represent a growing threat. Potential targets of agroterrorists employing biological agents could run the gamut of the farm-to-fork spectrum, including field crops; farm animals; animal feed; food items in the processing, transportation, and distribution chain; and market-ready foods. Even agricultural workers could be targets. Targets could be exposed to the pathogen by direct contact with the animal, crop, feed, or food item; by release of the agent into farm facilities and fields; or by aerosolizing the agent.

A Brief History

Biological agents have been an important component of warfare and terrorism for millennia. Examples include:

- Roman soldiers dumped bodies into the wells of their enemies to contaminate their water supplies; Assyrians poisoned their enemies' wells with toxic rye ergot; and Greeks and Carthaginians used plant toxins to poison water and wine.[1,2]
- German intelligence agents deliberately infected U.S. and allied draft horses and livestock with glanders and anthrax during World War I.[13,14]
- In 1952, a MauMau attack on Kenyan mission cattle used a local plant toxin.[13,14]
- Cattle were poisoned with anthrax by Rhodesian security forces in South Africa in the late 1970s.[13,14]
- Horses were infected with glanders by Russian security forces in Afghanistan in the early 1980s.[13,14]
- In 1984, a religious cult (Bhagwan Shree Rajneesh) deliberately contaminated an Oregon salad bar with salmonella in an attempt to influence turnout for a local election; 751 people subsequently got sick.[2,15]
- During World War II, Germany was suspected of dropping Colorado potato beetles over England, and Japan was accused of introducing anthrax and rinderpest into Russia and Mongolia.

After World War II, more than two dozen nations either acknowledged or were suspected of conducting offensive biowarfare programs.[1] U.S., U.K. and

U.S.S.R. programs included work on anti-agriculture agents. The U.S. has since learned that the bioweapons program in the U.S.S.R. employed 60,000 to 70,000 people, of whom some 10,000 were focused on anticrop and anti-livestock weapons.[16] By executive order, President Nixon terminated the U.S. offensive bioweapons program in 1972. Subsequently, the Biological and Toxin Weapons Convention of 1972 was signed by the U.S., the U.S.S.R., and 100 other countries.[17] Still, concerns persist that some nations continue bioweapons research to this day.

There are few documented cases of bioterrorism. Seth Carus of the National Defense University cited 222 bioterrorism-related incidents (including threats) during the 20th century.[13] Of these, only 24 were confirmed attacks; 11 involved food poisoning, and only 3 were directed against farm animals or crops.

Agroterrorism: How Real is the Threat?

Ominous Signs

The figures cited in the previous section should not give us a false sense of security. The last decade of the 20th century saw a significant increase in bioterrorism-related incidents; two-thirds of the total number of reported incidents in the century occurred in the 1990s. While this may reflect improved incident tracking and reporting procedures, FBI statistics indicated a rapidly growing interest by terrorists or criminals in the use of biological agents.[1,18] The ongoing threat of agroterrorism is underscored by the discovery of Al Qaeda's apparent interest in attacking U.S. agriculture. This is not surprising, given Osama bin Laden's overriding goal of inflicting major economic damage against the U.S.

In addition, the threat of terrorism directed against agricultural interests by extreme ideological groups is increasing. One interest group in particular, animal rights advocates, bears special mention. An investigation by the U.S. Department of Justice found that, between 1977 and 1993, there were 313 separate incidents involving acts of terrorism directed against animal enterprises by animal rights extremists.[19] While attacks against animal holding facilities may be conceived to protect animals from exploitation, the consequences can be harmful and even deadly to both animals and humans.

Furthermore, recent major incidents of foreign animal disease outbreaks and widespread food contamination incidents, in the U.S. and internationally, remind us of the relative ease and catastrophic consequences of the deliberate introduction of a pathogen at a critical node in the U.S. agriculture infrastructure.

Vulnerability of U.S. Agriculture to Disruption

The U.S. is exceedingly vulnerable to agroterrorism. Agroterrorism events directed against the U.S. could deliberately introduce livestock and poultry diseases or plant pathogens, intentionally contaminate human food or animal feed, or attack the physical infrastructure. This vulnerability was publicly stated by T. Thompson, Secretary of Health and Human Services, in December 2004: "For the life of me I cannot understand why the terrorists have not attacked our food supply because it is so easy to do."[20] The U.S. is vulnerable for several reasons:

- Americans take food for granted. U.S. citizens and residents take for granted the availability of abundant, affordable, and safe food but are largely ignorant about what it takes to produce, supply, and protect it. Terrorists, recognizing this complacency and ignorance, could exploit that vulnerability, leading to loss of confidence in the food supply, agricultural infrastructure, and government institutions, with significant economic consequences.
- U.S. agriculture is a tempting target. The vast size, complexity, economic value, structure, openness, and pervasiveness of the U.S. food and agriculture system make it a highly tempting. The target is particularly appealing to inimical interests who seek to undermine the nation's economic strength.
- The U.S. food and agriculture system is a soft target. The U.S. agricultural system is so open and ubiquitous that determined terrorists would have relatively easy access to the nation's farms, facilities, commodities, and food. Only in the last decade has the country begun to get serious about protecting its food and agriculture assets from deliberate disruption.
- The U.S. agricultural industry is highly concentrated. The high concentration of U.S. agriculture allows for the easy introduction of a pathogen or contaminant that can spread rapidly through a population of animals or plants or through postharvest foods in the processing, distribution, or sales chains.
- There is substantial movement of many agricultural products and commodities throughout the country. The widespread and often rapid movement of live farm animals and agricultural products between regions significantly increases the potential to quickly spread disease or contamination over a large area.
- Much of the industry is vertically integrated. This characteristic means that a contaminant or pathogen introduced at one point of a company's "farm-to-fork" continuum could rapidly spread (and multiply) throughout the continuum.

- U.S. farm animals and crops are especially susceptible to exotic diseases. Many exotic plant and animal diseases that are scourges in other parts of the world are not endemic in the U.S. Introduction of diseases to which the crops or livestock have not been exposed could have catastrophic consequences. Because of the relative health of U.S. farm species, vaccines are not routinely used, and there is a paucity of vaccines available to protect against foreign animal disease and exotic plant pathogens. There is limited genetic diversity in many U.S. agriculture species (several crops are virtual monocultures), resulting in increased susceptibility to diseases for which there is no inherent resistance. Animal husbandry practices (particularly crowding) can increase stress levels, also lowering resistance to disease. Widespread use of antibiotics to treat common livestock diseases can result in antibiotic-resistant strains of harmful bacteria.
- The U.S. food and agriculture industry employs numerous transient or undocumented workers. This increases the risk of deliberate disruption perpetrated by hostile or disgruntled employees and makes it difficult to determine the origin of an act of terrorism.
- There are few suppliers of some critical agricultural components. For example, there are only two U.S. manufacturers of the plastic material used in the production of milk bottles. One is in New Orleans. When Hurricane Katrina ravaged the city and shut down the production plant in 2005, the nation came perilously close to a major disruption of the $35 billion dairy industry.[6]
- It would be difficult to trace infected animals in the event of a disease outbreak. Efforts to establish and implement a National Animal Identification System were resisted by the livestock industry. A voluntary Animal Disease Traceability Framework is now under development and, if widely adopted, will help determine the origin and aid in understanding of the epidemiology of a disease outbreak.
- Our borders are porous. With today's limited commitments of resources and personnel to border security, it is virtually impossible to effectively monitor and interdict the movement of would-be terrorists and illegal or harmful materials crossing our borders by land, sea, or air.
- Large numbers of exotic animals are brought into the U.S. Every year the U.S. legally imports four million exotic birds, 640,000 reptiles, and 40,000 primates, most of which are designated for the pet industry.[21] There is significant potential for the accidental or deliberate introduction of exotic diseases from these animals.
- The U.S. imports substantial quantities of food and feed. The average American consumes 358 pounds of imported food products every year, and imported food—from more than 50 countries—comprises

17 percent of the American diet, including 60 percent of its fresh fruit and vegetables, 85 percent of its seafood, and nearly 60 percent of its spices.[22,23] The value of food imports is now $80 billion a year and continues to steadily increase. To supply the growing demand, nine million shipments annually enter 361 U.S. ports.[23] Only about one percent of these imports is physically inspected. Furthermore, much of the food we import comes from countries with which we do not have good relations. There is significant potential for the deliberate or accidental contamination of foods imported from other countries.

- Our knowledge base is insufficient. Our insufficient knowledge of factors affecting the transmission, spread, and potential control of animal, plant, and food pathogens and of the diseases that they cause represents a significant vulnerability to the deliberate introduction of infectious agents.

Potential Consequences of a Major Disruption to Food and Agriculture

Agroterrorism: Low Probability, High Consequence

Because there have been relatively few documented cases of bioterrorism and agroterrorism, and because a deliberate attack on U.S. agriculture interests would not have the immediate shock value of a highly visible event with mass human casualties, an agroterrorism incident may be a low-probability event. But the nation is exceedingly vulnerable to an attack on its agricultural infrastructure, and a successful attack would have devastating consequences. Any measure to deter, prevent, or effectively respond to an agroterrorism event will also strengthen our ability to address an unintentional outbreak of an agricultural disease or a major food contamination incident.

It is instructive to examine some case histories of events that led to major disruptions to the food and agriculture sector in this country and overseas. We will look at a catastrophic livestock disease outbreak, two cases involving introductions of exotic plant diseases, and a devastating food safety incident.

Case History: Foot and Mouth Disease, Taiwan, 1997

Foot and mouth disease (FMD) is a viral affliction that is considered to be the most contagious known disease affecting animals. It affects over 70 species of split-toed mammals, including swine, cattle, sheep, and goats. The disease exists in two-thirds of the world and is endemic in much of Asia, the Middle East, Africa, and parts of South America. Less than one percent of infected adult animals die from FMD; however, morbidity, manifested by suppressed appetite, lameness, fever, blisters, and open sores, is nearly 100 percent, rendering diseased livestock unfit for commerce. Further, FMD is very difficult to control once an outbreak occurs.

The United States has not had a confirmed case of FMD since 1929; however, recent and ongoing FMD outbreaks in other parts of the world give cause for great concern in this country. The best documented outbreak occurred in England in 2001. On February 23 of that year the disease was diagnosed at a farm not far from London. Within months, it had spread throughout England, Scotland, and Wales. Before it was contained, over six million animals had to be slaughtered, and losses to the United Kingdom's agriculture and tourism economies totaled $10 to 20 billion dollars.[24-26]

Less well known, but fraught with lessons for the United States, is an FMD outbreak that occurred in Taiwan in 1997. Before the disease struck, Taiwan was the world's third largest pork exporter, sending nearly a third of its production to overseas markets, with 95 percent of the exports going to Japan. Prior to the outbreak, 10 million hogs populated the island nation.[27]

In retrospect, the Taiwan FMD outbreak was an accident waiting to happen. The country produced little of its own hog feed and had to import it from other countries, driving up production prices. To expand the highly profitable export market, domestic prices were kept artificially high. This resulted in a substantial black market for smuggled pork products and hogs, much of it coming from mainland China. In addition, swine density was high; hog farms were located near slaughterhouses; no FMD vaccination program was in place; diagnostic lab capabilities were limited; the practice of garbage feeding was widespread; and there was substantial human social traffic among farms. Further complicating matters, swine vesicular disease, whose symptoms are nearly identical to those of FMD, is endemic to Taiwan, making rapid diagnosis difficult.[24,27]

On March 20, 1997, the Taiwan Council of Agriculture announced an FMD outbreak near Taipei. The government reacted as quickly as possible: they placed an immediate ban on pork exports, slaughtered hogs in infected herds (with military assistance), and initiated a vaccination program. Within a week of the announcement, the outbreak had spread to over 700 farms, infecting some 150,000 pigs. By mid-May the entire island nation was affected. By late summer the vaccination program finally halted the disease spread—over 13 million doses of a vaccine effective against the identified virus strain were eventually administered. But at the start of the outbreak only 40,000 vaccine doses were on hand, and internal controversies and indecision delayed the purchase of additional vaccine from overseas.[27]

The consequences of the outbreak were devastating. Nearly four million hogs—more than one-third of the nation's swine herd—were slaughtered, and 6,000 pig farms were completely depopulated. Export markets were wiped out almost overnight, and domestic pork prices crashed. Total losses to the economy were nearly $7 billion. Taiwan made indemnification payments to hog farmers who suffered losses, but the payments often exceeded the market value of the pigs, leading to accusations that some farmers deliberately infected their herds.[24]

After confirmation that the virus strain also existed in mainland China, it was considered likely that the origin of the virus was probably piglets or meat products smuggled in from China.[24,27]

Case History: Asian Soybean Rust, U.S., 2004

Soybeans are a vitally important U.S. agricultural crop, second only to corn in value and production. Annual cash receipts from soybean sales exceed $33 billion, nearly equivalent to that of the entire poultry industry.[28] More than one-half of U.S. states derive substantial revenues from soybean production, and soybeans are an important export crop.[29]

Asian soybean rust (ASR) is a fungal disease affecting a variety of commercial plant species, including soybeans. Two different species of the fungal genus, *Phakopsora*, are implicated in the disease, but only one, *P. pachyrhizi*, is highly virulent. First discovered in Japan in 1902, the disease spread worldwide over the next century. The rust was detected in Argentina in 2000. As U.S. scientists monitored its advance toward North America, they concluded that ASR would inevitably reach its coast, most likely between 2004 and 2007. Accordingly, they began to prepare.

Preparations included the establishment, in 2002, of an ad hoc working group under the leadership of USDA. Participants included several USDA agencies, the National Plant Board, a number of land grant universities, state agencies, and farm industry representatives. The working group identified and began to implement a four-pronged strategy to address the inevitable introduction of the disease:

1. Protection: identify potential pathways for introduction of ASR into the U.S. and prevent or slow the arrival of the pathogen
2. Detection: enable early detection and identification of ASR through improved education and diagnostics
3. Response: establish communications networks to share information and recommendations

4. Recovery: support research to develop improved fungicides and disease-resistant cultivars

Supplementing the working group's efforts, a newly established National Plant Diagnostic Network (NPDN) conducted 22 multi-agency exercises in 34 states to enhance the ability to rapidly detect the agent and to facilitate communications and strengthen relationships among agencies in the event that ASR entered the U.S.[30]

Not surprisingly, ASR was discovered in a field in Louisiana on November 9, 2004. It was detected quickly, largely due to the NPDN's training exercises.[31] Within three weeks the pathogen had spread to 8 more states in the southern U.S.[29] Since then, ASR has been detected in 16 states, including all of the southeastern U.S. and much of the Midwest. Nonetheless, the economic impact of the disease has been less than feared. This may be attributable to disease management approaches employed by soybean growers and partners (including the research community) that have built on the four-pronged strategy. The approaches include effective use of selected fungicides, implementation of a disease scouting/detection early-warning system that employs sentinel plots, development of simulation models for spore dispersion that help to predict outbreaks in advance, and cooperative research to improve understanding of the disease and potential control strategies.[32]

Case History: Citrus Canker, Florida, 1995–2006

Based on the latest statistics (circa 2007–2008), the economic impact of the citrus industry in Florida is approximately $9 billion annually. The industry accounts for 76,000 full- and part-time jobs in the state and brings in nearly 30 percent of all Florida farm receipts.[33] The Florida citrus industry is, however, under threat from two major diseases. One, Huanglongbing, or citrus greening, is a vector-borne bacterial disease transmitted by the Asian citrus psyllid, an invasive insect. First detected in Florida in 2005, the disease primarily affects the orange juice market. The economic impact has been substantial—over $3.6 billion in lost revenue and more than 6,600 lost jobs.[34]

The other disease, citrus canker, has had a much longer history in Florida and was first detected in 1912. The introduction was attributed to citrus tree seedlings imported from Japan. The state responded aggressively, and the disease was declared eradicated by 1933. It emerged again in 1986 in Manatee County and was considered to be eradicated by 1994, but was discovered again in the same area three years later. In the meantime, citrus canker was detected in urban Miami in 1995. The disease began to spread.[35,36]

In response to the Miami-area outbreak, authorities established a cooperative state-federal citrus canker eradication program. The program focused on southeastern Florida and then expanded to Manatee County. The initial response was to remove all citrus trees within 125 feet of an infected tree. When USDA epidemiological studies determined that the pathogen could spread much further—especially in rainstorms accompanied by wind—scientists, regulators, and citrus growers settled on a 1,900-foot radius as the potential infection zone. This became the basis of a state policy, implemented in 2000, that called for mandatory eradication of all citrus trees (both commercial and "dooryard") within that distance of an infected tree. The policy included federal quarantine boundaries (approximately two miles around infected trees), within which movement of citrus plant material was restricted, and decontamination of all persons and vehicles entering or leaving the quarantine zone.[36,37]

By March 2002, over one and a half million commercial citrus trees and 600,000 dooryard trees had been removed or pruned. But the eradication program did not appease homeowners whose dooryard trees lay within an infected zone. They brought suit in Broward County in 2002, with the result that the eradication program was put on hold while the suit made its way through the courts. Hurricanes in 2004 and 2005 spread the disease still further, with 24 Florida counties infected with canker by 2005; the disease was detected even in a citrus nursery. Inevitably, Florida ended its mandatory eradication program on January 11, 2006, though voluntary tree removal is still encouraged, decontamination policies remain in effect, and the entire state of Florida is under quarantine for the disease.[35]

Case History: Melamine Contamination, 2007–2008

On March 15, 2007, the Food and Drug Administration (FDA) confirmed that unexplained widespread deaths of cats and dogs were attributable to contaminated pet feed. Further investigation revealed that the contaminant was melamine, a man-made, nitrogen-rich chemical normally used in the manufacture of plastics and fertilizer. The FDA traced the feed and contaminant to China and concluded Chinese pet food manufacturers had deliberately added melamine to feed ingredients, including wheat gluten and other vegetable proteins; under the guise of elevated protein content. Pets consuming the contaminated feed experienced acute renal failure.[38,39,40]

Sixty percent of American households own pets, and the melamine contamination incident soon took on national security implications, fully engaging the Department of Homeland Security.[41] FDA responded to hundreds of calls daily from distraught or anxious pet owners, voicing complaints and demanding action.[42]

The federal response was rapid, aggressive, and well-coordinated. To manage information flow and reporting, DHS established an interagency working group (IAG) under its Office of Health Affairs (OHA). The department also established a formalized process for coordinating the interagency response—the first time such a process had ever been applied to a food safety incident.[41]

FDA quickly traced the origin of the contaminated feed ingredients from Chinese suppliers through U.S. importers to feed ingredient and pet food manufacturers in the U.S. and Canada and then to feed mills, hog and poultry farms, fish hatcheries, and aquaculture facilities. Beginning on April 18, 2007, the agency initiated the largest recall in its history to date and, with the cooperation of state officials, conducted on-site investigations of manufacturing, warehouse, and feed distribution sites. Scientists across the country collaborated to investigate the specific mechanisms of melamine toxicity.[42,43]

An interagency risk assessment concluded that the risk to human health was negligible.[41,42] On September 9, 2008, the *Shanghai Daily* reported that 14 children from one Chinese province suffered from kidney failure after consuming a specific brand of powdered infant milk formula. Chinese authorities seized and recalled over 10,000 tons of milk powder. Further investigation revealed that 22 separate companies sold contaminated powdered infant milk formula. Over 300,000 Chinese infants were affected, and six children died. As the investigation continued, it became apparent that a large variety of dairy and nondairy products from China contained melamine.[42,44] In November 2008 FDA announced that it found trace levels of melamine in one sample of infant formula made in the U.S. and in a few samples of nutritional and medical supplements; the agency concluded the levels were too low to be a health risk.[45]

The political and legal response to the melamine contamination episode was strong. The U.S. Congress passed legislation (Title 10 of the FDA Amendments Act) strengthening ingredients, processing, and labeling standards; called for early warning surveillance and notification of veterinarians and stakeholders during recalls; and required a searchable database of recalled pet foods and a reportable food registry for foods having a reasonable probability of affecting the health of humans or animals.[42] In China the response was even more severe. By November 2009, Chinese authorities had arrested 21 food industry executives and managers and executed two of them. But it was too late to stop the worldwide distribution of contaminated products that, by 2009, had spread throughout Russia, through much of Southeast Asia and Europe, to Australia and to several African nations.[44]

Lessons Learned

Although none of the incidents discussed above involved terrorism, each has lessons that are applicable to the biosecurity of food and agriculture and to defending the U.S. infrastructure against both unintentional and deliberate attacks.

The FMD outbreak in Taiwan demonstrated the inadequate preparation of Taiwanese farmers and government officials for the introduction of the disease; poor biosecurity practices that increased the likelihood of disease introduction and the potential for disease spread; and an ill-considered national policy response that exacerbated the spread of the disease and the magnitude of losses:

The introduction and spread of Asian soybean rust in the U.S. illuminates different, more positive lessons for the future. In this case, we had anticipated for years that the disease would hit our shores, and we had prepared by developing a coordinated, four-pronged strategy of protection, detection, response, and recovery. When the pathogen arrived, we immediately implemented the strategy, most likely reducing the spread and impact of the disease. Still, there are two areas that require significantly greater attention if we are to improve management of the disease in the future. The first is expanding education about the disease and, specifically, about when and how to apply fungicides to growers. The second is increasing research to develop, evaluate, and establish ASR-resistant soybean strains.

The response to the late 20th century citrus canker outbreak in Florida was timely, aggressive, well-founded in science, and—based on a previous history of effective containment of the disease—had great promise for success. Yet the response failed: the outbreak continued to spread, and the state ultimately abandoned its mandatory program to eradicate infected and exposed citrus trees. The primary reason for the failure was sociopolitical. State and federal authorities did not adequately consider the public reaction to the forced removal of dooryard citrus trees and did not anticipate the interruptions in the eradication program engendered by the lawsuits. Although the hurricanes of 2004 and 2005 resulted in further spread of the disease, we may never know whether an uninterrupted eradication program might have eliminated the disease before the storms hit.

The U.S. response to the melamine episode is a model of how to manage a food safety crisis. The establishment of an IWG and a formalized mechanism for response enabled an effective, coordinated process for managing and disseminating information, conducting critical research, and developing and implementing appropriate public policy, including widespread recalls and relevant legislation. Even so, the episode underscored a key U.S. vulnerability in food and agriculture biosecurity—a woefully inadequate program for monitoring and inspecting imports. The melamine incident elucidated the consequences of deliberate attacks on agricultural products. To summarize, there are three key lessons that relate to agroterrorism:

- It is highly likely that exotic plant diseases could contaminate our agricultural system. As a country we need to focus on containment and recovery efforts after a suspected attack.
- Research needs to be expanded; focusing on improved detection and diagnosis, epidemiology, more effective vaccines and therapeutics, on-farm management practices, and development of disease-resistant crops and livestock.
- We must begin to understand that effective biosecurity is not only a function of best management practices but is also dependent on sociopolitical considerations.

FOOD DEFENSE: PROTECTING U.S. FOOD AND AGRICULTURE FROM INTENTIONAL DISRUPTION

U.S. Food Defense since 9/11: Overview

Prior to the attacks of 9/11, Americans gave little thought to the potential of agroterrorism. This was well exemplified by a General Accounting Office report on combating terrorism issued nine days after the attacks. According to the agency, the report "did not focus on terrorism directed against agricultural targets" because agriculture was not considered a critical national infrastructure."[1]

As a result, the nation was ill prepared to address the threat and consequences of a deliberate attack on the food and agriculture infrastructure. The regulatory framework was inadequate, with no laws specifically addressing agroterrorism. Federal leadership related to food defense was fragmented, with many overlapping jurisdictions: policies and responsibilities were spread among seven federal agencies and more than 30 different programs, and there was little coordination among them. No single agency was designated to have responsibility for overall leadership and coordination. The U.S. Department of Agriculture

(USDA) was not a significant player in the national security arena, and few USDA personnel had the requisite security clearances. Linkages to state and local authorities and to the private sector were weak and tenuous.[1]

In the aftermath of the 9/11 attacks, it became increasingly clear that a key Al Qaeda goal was to damage the U.S. economy. Based on the economic value of the U.S. food and agriculture sector and the sector's vulnerability to deliberate attacks, federal authorities included agroterrorism in the threat matrix to be addressed. In the next several years a number of public regulatory, policy, and program initiatives in support of food defense underscored this new focus on agriculture. At the same time the private sector, often in partnership with government, strengthened its capacity to address the threat of agroterrorism by improving biosecurity and biosafety in the industry.

U.S. Food Defense: Looking to the Future

Gaps and Recommendations

As discussed in Addendum 8.1, the U.S. has made substantial progress in developing and implementing programs and policies to protect the food and agriculture infrastructure against deliberate disruption (Table 8.1). Most of these measures simultaneously help protect the sector against unintentional disruption.

Nonetheless, the nation continues to face a number of stiff challenges as it defends its food and agriculture sector. These challenges fall primarily under the headings of a national regulatory framework, leadership and coordination, and federal funding support.

Despite the substantial food defense-related federal regulatory framework established after 9/11, there is still need for additional clarity related to the roles and responsibilities of federal agencies in general and food safety agencies in particular. USDA-FSIS and HHS-FDA share regulatory responsibilities for food safety, but the division of responsibilities defies logic, reflects historical biases and precedents that are no longer relevant, and confuses efforts to protect the postharvest food supply. For example, FSIS has regulatory responsibility for meat; poultry; frozen, dried, and liquid eggs; and catfish, while FDA has regulatory responsibility for nearly everything else (amounting to nearly 80 percent of the U.S. food supply). FDA's responsibilities include oversight of dairy products, plant products, spices, grain-based foods, and all seafood except catfish. Complicating matters further, the food safety-related provisions of the

TABLE 8.1 Comparison of U.S. Food Defense Preparedness Before and After 9/11/2001

Before 9/11	After 9/11
Agriculture not a critical national infrastructure	Agriculture a critical national infrastructure
Limited awareness; human health focus	More attention to agriculture
Regulatory framework inadequate	Substantial regulatory framework
USDA not a national security player	USDA at the table
Many overlapping jurisdictions	More consolidation
Poor coordination among federal agencies	Better coordination (HSPD-7,-9)*
Poor linkages to state/local authorities	Better linkages*
Private sector insufficiently involved	Private sector more involved*
Significantly underfunded	Increased funding
No one "in charge"	DHS leadership role

*e.g. Food and Agriculture Sector Coordinating Councils (government and private sector); Strategic Partnership Program Agroterrorism (USDA, FDA, DHS, FBI with states and private sector: assessments of infrastructure).

Bioterrorism Act of 2002 and the Food Safety Modernization Act of 2011 apply only to FDA. FSIS and FDA have established a close working relationship and have collaborated effectively with each other and with the private sector on food defense initiatives since 9/11. Nonetheless, consolidation of federal food safety responsibilities within a single agency could enhance the nation's food defense preparedness and response programs. Appendix 1 of this book contains a summary of policy initiatives that supplement this discussion.

Prior to 9/11, federal responsibilities for food defense were spread among dozens of programs and agencies, and no single agency was designated to provide overall leadership or coordination.[1] Coordination between the federal government and the private sector was limited and often ineffectual. Today, DHS has the lead responsibility for coordinating counterterrorism programs, including agroterrorism, and the National Response Framework designates USDA as the sector specific agency with lead responsibility for the protection of food and agriculture. Coordination and cooperation between government and the private sector are substantially improved, as discussed earlier in this chapter.

However, the number of federal agencies and programs with responsibilities related to food defense is at least as large as it was before 9/11, and many food and agriculture companies have still not developed and implemented effective biosecurity and food defense plans. There remains a need for substantially improved coordination among federal agencies (including consolidation or elimination of competing or overlapping programs and responsibilities), and for more effective engagement of the private sector in food defense planning.

Federal Support for Food Defense

As discussed earlier in this chapter, it is difficult to tease out federal funding support specifically targeted for food defense from that appropriated for the broader category of biodefense. What is clear is that appropriations for biodefense programs in USDA have remained at about $100 million annually for the past three fiscal years. This is less than half the funding level for the five preceding fiscal years (FY 2005-FY 2009) and is approximately two percent of the funding appropriations for biodefense for the Department of Health and Human Services.[46]

Funding agricultural research is a good investment, in terms of both the national economy and national security. The average annual economic rate of return on investment to the public is 53 percent, meaning that an initial $100,000 investment in agricultural research will yield annual benefits of $53,000 for several decades.[47] Agricultural research in animal health, plant diseases, and food safety not only helps protect the nation against deliberate attacks on the food and agriculture infrastructure but also makes the U.S. more resistant and resilient to natural disease outbreaks or food contamination incidents. Equally important, research in support of a vibrant agricultural economy helps ensure long-term food security for the U.S. Given these considerations, it might seem logical that there would be strong support for agricultural research; however, research funding has been declining. If the U.S. is to maintain its global supremacy in food production and ensure protection of its food and agriculture resources against deliberate or unintentional disruption, it is critical that investment in agricultural research be significantly expanded. Research should address the priorities identified in the agriculture and food sector specific plan, discussed above.

Federal support for food defense should also be expanded for outreach and professional development (including education and training programs for large animal veterinarians, food defense exercises involving multiple stakeholders, and public education to help citizens understand the importance of, and approaches to, protecting the food and agriculture infrastructure); for the National Plant Diagnostic Network and the National Animal Health Laboratory Network; for the National Plant Disease Recovery Program; and for the National Veterinary Stockpile. All of these programs are critical for U.S. food defense but are funded at levels well below that needed for effective implementation.

Resilience-Based Strategies for Food Defense

As illuminated by the case histories discussed earlier in this chapter, even the most concerted efforts to prevent the introduction and spread of animal or plant diseases or a major food contamination event may fail. That is why strategies to manage and respond to such events are as important as prevention strategies in a comprehensive food defense plan. We must assume that major disease or food contamination episodes will occur and prepare for them accordingly in food defense planning.

Strategies based on resilience should be a key component of food defense plans. Resilience has been defined as "the ability to recover readily from illness, depression, adversity, or the like."[48] Resilience in the food and agriculture sector may be defined as "the ability to quickly restore food security for affected Americans, and economic security for the food and agriculture sector, after a major disruption event." The goals of resilience-based strategies in agriculture should be to reestablish production, supplies, and distribution of affected food items to pre-event levels as soon as possible, to ensure adequate availability of acceptable substitutes during the recovery period, and to minimize the economic impact on the agriculture sector. Resilience strategies should be based on the ability to rapidly assemble and deploy adequate supplies of germplasm and seedstock (including of disease-resistant varieties) to repopulate lost livestock and crops and to secure and distribute supplies of affected food items, or acceptable alternatives, from other production areas or storage facilities in the U.S. and overseas;

To achieve resilience in food and agriculture, the U.S. should accomplish the following:

- Expand genetic diversity among animal and plant species important for agriculture. The U.S. presently relies too much on relatively few genetic strains of important livestock, poultry, and crop species. Greater genetic diversity will reduce vulnerability to specific pathogens.
- Increase efforts for selective breeding or genetic manipulation to improve resistance to specific diseases.
- Significantly expand seed banks and germplasm repositories to enable rapid repopulation of livestock or crops lost to disease outbreaks, including expanded support for the National Plant Disease Recovery System.
- Modify and expand federal disaster relief programs and loss compensation programs to reduce risks for farmers affected by agroterrorism. Presently these programs are overwhelmingly focused on natural disasters.[49]

- Reduce the heavy concentrations of livestock, crops, and products in relatively confined areas, where introduced diseases will spread rapidly.
- Provide incentives for farmers to rotate crops more frequently.
- Provide additional incentives and support for small farmers with local or regional markets. In the event of significant disruption to large farms in major food-producing regions, local food sources may be vitally important.
- Expand long-term storage capacity for food commodities that can be made readily available to offset food shortages in an emergency. For decades the U.S. has operated a program to purchase surplus commodities from farmers to maintain price levels. The surplus food, temporarily stored in warehouses, is made available to domestic nutrition programs for low-income Americans and exported to other nations experiencing food shortages. In recent years, rising prices of commodities have sharply reduced U.S. food surpluses, with the result that "USDA has almost no extra food to supplement the billions in cash payments it spends to combat hunger at home and in developing nations."[50] In raw economic terms, government purchases of commodities for warehouse storage may not make sense; however, they could be an important component of food defense and food security in the U.S.
- Educate consumers about alternative food choices that could be nutritious, affordable substitutes for more traditional foods whose supply would be affected by a major disruption event.

CONCLUSION

The U.S. food and agriculture sector is a highly successful enterprise, a critical national infrastructure, and a key component of the national economy. It is also very vulnerable to deliberate or unintentional disruption. A major outbreak of a disease of livestock or crops or a significant food contamination episode could have devastating consequences for Americans and could severely impact national security.

Protection of the nation's food and agriculture infrastructure against disruption requires a strong food defense program that embodies elements of both biosecurity and biosafety. Since the attacks of 9/11, the U.S. has made substantial progress in developing and implementing food defense strategies in the public and private sectors.

But significant challenges remain. In particular, there is a need for greater clarity in the national regulatory framework governing food safety. Federal leadership and coordination of food defense remains fragmented, with too many overlapping or competing programs and responsibilities. Federal support for food defense programs, especially funding for agricultural research, is inadequate.

Finally, the development and implementation of resilience-based strategies for food defense will strengthen the nation's capacity to weather a major disruption to its food and agriculture sector.

References

1. Parker, Henry S. Agricultural Bioterrorism: A Federal Strategy to Meet the Threat. Institute for National Strategic Studies, National Defense University 2002.
2. Olson, D. Agroterrorism: Threats to America's Economy and Food Supply. FBI Law Enforcement Bulletin (February 2012). Available at: http://www.fbi.gov/stats-services/publications/law-enforcement-bulletin/february-2012/agroterrorism.
3. United Nations Food and Agriculture Organization. World Food Summit, 13–17 November 1996. Rome Declaration on World Food Security and World Food Summit Plan of Action. Available at: http://www.fao.org/docrep/003/w3613e/w3613e00.htm.
4. U.S. Department of Agriculture, Economic Research Service. Agricultural Outlook Statistical Indicators, Key Statistical Indicators of the Food and Fiber Sector (data as of January, 2012). Available at: http://www.ers.usda.gov/data-products/agricultural-outlook-statistical-indicators.aspx.
5. Capeheart, T., Allen, E., and Bond., J.K. Feed Outlook. United States Department of Agriculture Economics Research Service Outlook No. FDS-12J (October 2012). Available at: http://www.ers.usda.gov/publications/fds-feed-outlook/fds-12j.aspx.
6. U.S. Department of Agriculture Economic Research Service. Effects of Trade on the U.S. Economy. 2012 Data Overview. Available at: http://www.ers.usda.gov/data-products/agricultural-trade-multipliers/effects-of-trade-on-the-us-economy.aspx.
7. U.S. Department of Agriculture. Economic Research Service. U.S. Agricultural Trade (2012). Available at: http://www.ers.usda.gov/topics/international-markets-trade/us-agricultural-trade.aspx.
8. U.S. Food and Drug Administration. Food Defense Acronyms, Abbreviations, and Definitions (2012a.). Available at: http://www.fda.gov/Food/FoodDefense/EducationOutreach/ucm296330.htm.
9. Kelley, L. Food Defense: Protecting the Food Supply. Presentation to Georgetown University Medical Center course in Agroterrorism (February 10, 2012).
10. Ramsey. W.L. A Fresh Look at Food Defense. Security Technology and Design, Volume 17, Number 6 (June 2007). Available at: http://arc.securityinfowatch.com/article/10517254/a-fresh-look-at-food-defense?page=4 http://news-business.vlex.com/vid/fresh-look-at-defense-64030891).
11. Douglass, J.D., and Livingstone, N.C. America the Vulnerable: The Threat of Chemical and Biological Warfare. Lexington Books (1987).
12. Pavlin, J.A., Eitzen, E.M., and Caudle, L.C. Biological Warfare Defense. In: Kelley, P. (editor). Military Preventive Medicine: Mobilization and Deployment, Volume I, Chapter 28, pp. 627–654 (2004).

13. Carus, S. Bioterrorism and Biocrimes: The Illicit Use of Biological Agents in the 20th Century. Center for Counterproliferation Research, National Defense University (July 1999 revision).
14. Wilson, T.M., Logan-Henfrey, L., Weller, R., and Kellman, B. Agroterrorism, Biological Crimes, and Biological Warfare. In: Brown, C., and Bolin, C. (eds.). Emerging Diseases of Animals, ASM Press, pp. 23–57. Washington, DC (2000).
15. Tucker, J.B. Historical Trends Related to Bioterrorism: An Empirical Analysis. Emerging Infectious Diseases. Special Issue, Volume 5-No. 4 (July-August 1999). Available at: ftp://ftp.cdc.gov/pub/eid/vol5no4/ascii/tucker.txt.
16. Tucker, J.B. Biological Weapons in the Former Soviet Union: An Interview with Dr. Kenneth Alibek conducted by Jonathan B. Tucker. The Nonproliferation Review (Spring-Summer 1999).
17. Parker, H.S. Protecting Food and Agriculture: Private Sector Perspectives. Presentation to Georgetown University Medical Center course in Agroterrorism (February 4, 2011).
18. Orenstein, J.B. Now Fear This. Washington Post (December 26, 1999).
19. U.S. Department of Justice. Report to Congress on the Extent of Domestic and International Terrorism in Animal Enterprises (September 2, 1993).
20. Branigan, W., Allen, M., and Mintz, J. Tommy Thompson Resigns from HHS. Washington Post (December 3, 2004).
21. Heckert, R. Agrosecurity Challenges in the 21st Century. Presentation to Georgetown University Medical Center course in Agroterrorism (February 24, 2012).
22. U.S. Department of Agriculture. Economic Research Service. U.S. Agricultural Trade: Import Share of Consumption (2012e). Available at: http://www.ers.usda.gov/topics/international-markets-trade/us-agricultural-trade/import-share-of-consumption.aspx.
23. Smith deWaal, C. Import Inspection Failures and What Must be Done. Testimony of Caroline Smith DeWaal, Director of Food Safety, Center for Science in the Public Interest, before the United States House of Representatives Energy and Commerce Committee's Subcommittee on Oversight and Investigations. Washington, DC (July 17, 2007). Available at: http://democrats.energycommerce.house.gov/Press_111/20090311/testimony_dewaal.pdf.
24. Extension Disaster Education Network, Agricultural Disasters: Foot and Mouth Disease (2011a). Available at: http://eden.lsu.edu/Topics/AgDisasters/FMD/Pages/default.aspx.
25. House of Commons Library. Foot and Mouth Disease. Research Paper 01/35, 27 (March 2001). Available at: http://www.parliament.uk/documents/commons/lib/research/rp2001/rp01-035.pdf.
26. National Cattlemen's Beef Association. FootAndMouthDiseaseInfo.org (2012). Available at: http://www.footandmouthdiseaseinfo.org/default.aspx.
27. U.S. Department of Agriculture. Foreign Agricultural Service. Foot-and-Mouth Disease Spreads Chaos in Pork Markets. Livestock and Poultry World Markets and Trade. FASonline (October, 1997). Available at: http://www.fas.usda.gov/dlp2/circular/1997/97-10LP/taiwanfmd.htm.
28. U.S. Department of Agriculture. Economic Research Service. Farm Income and Wealth Statistics: United States: Leading commodities for cash receipts 2011 (2012d). Available at: http://www.ers.usda.gov/data-products/farm-income-and-wealth-statistics.aspx#27415.
29. Schnepf, R. Asian Soybean Rust Background and Issues. Congressional Research Service, CRS Report for Congress (updated January 12, 2005). Available at: http://www.nationalaglawcenter.org/assets/crs/RL32225.pdf.

30. National Plant Diagnostic Network. Welcome to the National Plant Diagnostic Network Website (2012). Available at: http://www.npdn.org/.
31. National Plant Diagnostic Network. Case Study: Asian Soybean Rust (undated). Available at: http://www.npdn.org/webfm_send/1710.
32. Doerge, T., and Trygom, J. Asian Soybean Rust in the U.S.: 2007 Experiences and Future Management. Crop Insights, Pioneer Agronomy Sciences. Volume 18, No. 4, 4 (2008). Available at: http://www.mccormickcompany.net/pioneer/cropinsights/74.pdf.
33. Florida Citrus Mutual. Citrus Statistics (2012). Available at: http://flcitrusmutual.com/citrus-101/citrusstatistics.aspx?q=economic+impact&submit.x=46&submit.y=9# .
34. University of Florida. Citrus greening costs $3.63 billion in lost revenues and 6,611 jobs, new UF study shows. University of Florida News (January 24, 2012). Available at: http://news.ufl.edu/2012/01/24/greening-cost/.
35. University of Florida, Citrus Canker: History (2012b). Available at: http://www.crec.ifas.ufl.edu/extension/canker/history.shtml.
36. Gottwald, T.R., Graham, J.H., and Schubert, T.S. Citrus Canker: The Pathogen and its Impact. APSnet, American Phytopathological Society (2002). Available at: http://www.apsnet.org/publications/apsnetfeatures/Pages/CitrusCanker.aspx.
37. University of Florida. Citrus Canker: Decontamination (2012c). Available at: http://www.crec.ifas.ufl.edu/extension/canker/decontamination.shtml.
38. Brown, D. How Two Innocuous Compounds Combined to Kill Pets. Washington Post (May 7, 2007). Available at: http://www.washingtonpost.com/wp-dyn/content/article/2007/05/06/AR2007050601034.html.
39. Weise, E., and Schmit, J. Melamine in pet food may not be accidental. USA Today (April 20, 2007). Available at: http://usatoday30.usatoday.com/money/industries/2007-04-19-pet-food-usat_N.htm.
40. U.S. Food and Drug Administration. Melamine Pet Food Recall of 2007 (2010). Available at: http://www.fda.gov/animalveterinary/safetyhealth/recallswithdrawals/ucm129575.htm.
41. McGinn, T. Food Defense: Post-Harvest Preparedness Statement for the Record before the U.S. House of Representatives Committee on Homeland Security Subcommittee on Emerging Threats, Cybersecurity, and Science & Technology (July 24, 2007). Available at: http://chsdemocrats.house.gov/SiteDocuments/20070724163233-37162.pdf.
42. Reimschuessel, R. Melamine: The Anatomy of an Adverse Event. Pet Food Recall of 2007 and the Infant Formula Incident in China 2008. PowerPoint presentation (2009). Available at: http://www.jifsan.umd.edu/docs/advisory2010/presentations/final/%281%29%20Reimschuessel.pdf.
43. Dobson, R.L.M., Motlagh, S., Quijano, M., Cambron, R.T., Baker, T.R., Pullen, A.M., Regg, B.T., Bigalow-Kern, A.S., Vennard, T., Fix, A., Reimschuessel, R., Overmann, G., Shan, Y., and Daston, G.P. Identification and Characterization of Toxicity of Contaminants in Pet Food Leading to an Outbreak of Renal Toxicity in Cats and Dogs. Toxicol. Sci. 106(1): 251–262 (2008). Available at: http://toxsci.oxfordjournals.org/content/106/1/251.full.
44. Gossner, C. M-E., Schlundt, J., Embarek, P.B., Hird, S., Lo-Fo-Wong, D., Beltran, J.J.O., Teoh, K.N., and Tritscher, A. The Melamine Incident: Implications for International Food and Feed Safety. Environ Health Perspect. 117(12): 1803–1808 (2009). Available at: http://www.ncbi.nlm.nih.gov/pmc/articles/PMC2799451/.

45. Zhang, J. and Wang, S.S. FDA Tests Find Melamine Traces in Baby Formula. Wall Street Journal (November 26, 2008). Available at: http://online.wsj.com/article/SB122764783105057435.html.
46. Franco, C. and Sell, T.K. Federal Agency Biodefense Funding. FY 2011-FY 2012. Biosecurity and Bioterrorism: Biodefense Strategy, Practice and Science. 9(2): 117–137 (2011). Available at: http://www.upmc-biosecurity.org/website/resources/publications/2011/pdf/2011-06-07-biodeffunds.pdf.
47. Farm Foundation. Issue report: Agricultural Research and Productivity for the Future (September 2009). Available at: http://www.farmfoundation.org/news/articlefiles/105-FINAL%20Sept09_IR%20(2).pdf.
48. Webster's Encyclopedic Unabridged Dictionary of the English Language, New Revised Edition. Gramercy Books, New York (1996).
49. U.S. Department of Agriculture. Farm Service Agency. Disaster Assistance Programs (2012). Available at: http://www.fsa.usda.gov/FSA/webapp?area=home&subject=diap&topic=landing.
50. Kirchoff, S. Surplus U.S. food supplies dry up. USA Today (May 2. 2008). Available at: http://usatoday30.usatoday.com/money/industries/food/2008-05-01-usda-food-supply_N.htm.

CHAPTER 9

Biosecurity in Mass Transportation Networks

Brian K. George

INTRODUCTION

Biosecurity in mass transportation networks (MTN) requires careful monitoring of high threat areas and coordination with local, state, and federal agencies to develop a concept of operations that integrates early warnings with well-coordinated response plans. The first portion of this chapter will center on information about mass transit agencies and their security needs and a general overview of the local, state, and federal government efforts to enhance biosecurity guidance. The second section will concentrate on the general characteristics of biological agents that a terrorist organization could potentially release in a mass transit facility, compare the advantages and disadvantages of existing biological detection systems, and address the Department of Homeland Security's action plan to acquire next generation systems. The major focus of this chapter is centered on security countermeasures: specific biological detection capabilities, policies, procedures, and operations that will help detect, react to, and respond to a biological agent release in a mass transit system.

MASS TRANSIT SYSTEMS AND VULNERABILITIES

Multimodal transit agencies usually consist of bus, light rail, commuter rail, heavy rail, or any combination thereof. A multimodal agency also contains administrative operations, and usually a senior management team or operating board makes decisions required for daily and annual activities. Most agencies possess a cadre of assets ranging from bus terminals, rail stations, and platforms to parking garages, maintenance facilities, and fuel storage units. These transit functions and their various assets make up a standard U.S. mass transit agency, and each component is included in the overall business plan. Most residents of a major city within the U.S. use or can use a multimodal transit agency. Millions of citizens use these systems every day to commute to work or while vacationing. Examples of multimodal transit agencies include the Port Authority of New York and New Jersey, South Florida Regional Transportation Authority, Detroit Transit Authority, Memphis Area Transit Authority, Dallas Area Rapid Transit, Denver Regional Transit Authority, and Los Angeles County Metropolitan Transportation Authority. Each agency possesses unique facets that assist or curtail their ability to service their patrons, and provide safe and secure rail and bus operations. After 9/11, each transit agency started to look at its existing operations in order to determine the potential terrorist threat, system shortfalls, and policy gaps that could reduce the overall ability to provide a safer environment.

Historically, transit agencies strive to keep trains, subways, and buses readily accessible, convenient, and inexpensive. As such, transportation systems are extremely susceptible to terrorist attacks. As terrorists continue to target public transportation worldwide, American transit agencies should be concerned. Since terrorists continue to use large-scale, indiscriminate violence, transit agencies must face the possibility that they might use small or large conventional explosives; a chemical, biological, or radiological dispersion device; or a nuclear weapon.[1] Recognizing the needs of transit agencies and jurisdictions to identify and prepare for potential terrorism risks, this chapter will attempt to develop a basic list of countermeasures to assist in reducing the vulnerability and consequences of a biological agent attack on a multimodal transportation system.

To attain a safer and more secure transit environment, each multimodal transit agency needs to conduct a risk-based assessment (similar to the threat analysis discussed in Chapter 6) that focuses on the transportation services provided. The chosen methodology needs to compare the relative risk that biological weapons (BW) or other weapons of mass destruction (WMD) will be used against critical assets owned or operated by the transit agency. This methodology should identify and prioritize potential security enhancements, emergency response plans, and recovery operations that an agency can implement to reduce those risks. Chemical, biological, radiological, and nuclear attacks remain extremely difficult for terrorist organizations to execute, but transit agencies must prepare to deal with these WMDs in order to reduce stakeholders' liability. The impact of such an attack will cripple the mass transit agency and most of the economic return the system provides to the greater metropolitan area.

It is important for transit agencies and jurisdictions to create the safest environment that their available resources allow. In order to achieve a safer transit environment, security planners and emergency managers must adopt and conduct an analytical process, similar to a risk assessment, to determine which security countermeasures will work best for each of their assets. The agreed process should allow each agency or jurisdiction the ability to customize its security countermeasures. In addition, the design must allow future enhancements to existing technologies to ensure that security planners and emergency managers use a similar baseline assessment. This adaptability is essential to ensure that the biological security system is up-to-date, as technology continues to rapidly evolve.

FEDERAL, STATE, AND LOCAL BIOSECURITY ACTIVITIES AND PLANNING

Before September 11, 2001, local fire and police departments created, exercised, and executed concepts of operations (CONOPS) for responding to incidents associated with fires and natural disasters affecting mass transit facilities. In areas of the country with high-use mass transit facilities, such as the northeast corridor or earthquake zones in western portions of the nation, local and state jurisdictions exercised mock response drills for natural disasters, but no agency properly prepared for a possible biological attack. However, the 2001 *Bacillus anthracis* (the causative agent of anthrax) attack on the offices of Senators Patrick Leahy and Tom Daschle caused 22 people to contract anthrax (11 by inhalation and 11 cutaneous), which resulted in the deaths of 5 individuals from inhalation anthrax. These events changed the preparedness landscape for local and state governments.[2] They forced local and state governments to develop specific CONOPS for biological agent incidents, especially in mass transit facilities. Unfortunately, the federal government lacks the coordination to ensure that local and state governments and transportation agencies receive general federal guidelines to develop transit security planning specific to biological attacks.

According to a 2007 congressional report (Detour Ahead: Critical Vulnerabilities in America's Rail and Mass Transit Security Programs) local and state agencies are responsible for the security and CONOPS development of their multimodal transit networks.[3] Further, the Transportation Security Agency (TSA) remains responsible for the development of a national strategy for transportation security, but it seems to focus most of its resources and staff on aviation security and not on rail or mass transit. To date, numerous threat and vulnerability analyses conducted by DHS and federal contractors failed to coordinate their assessment methods or to unify their security countermeasure recommendations to uniformly enhance transit security in the face of conventional and WMD attacks. In addition, the Homeland Security Act designates the Department of Transportation (DOT) responsible for rail safety and security and for safeguarding all shipments of hazardous materials.[4] Therefore, confusion abounds, and local and regional transit agencies remain responsible for bridging the gap in providing a safe environment for their patrons.

The substantial lack of unified or even coordinated federal guidance leaves the local, state, and private stakeholders responsible for developing their WMD CONOPS. This situation forces local and private/semiprivate transit agencies to maximize their limited resources and available subject matter expertise in developing biological agent response plans and CONOPS to deal with enhancing their security, responding to an event, and recovering from a clandestine release of an agent.[5] To attain a safer and more secure transit environment, each multimodal transit agency needs to conduct a risk assessment that focuses on the transportation services provided with its particular assets. The chosen methodology needs to compare the relative risk of acts of terrorism against critical assets owned or operated by the transit agency. This methodology should identify and prioritize potential biosecurity enhancements, emergency response plans, and recovery operations that an agency can implement to reduce those risks.

Harmful Biological Agents Overview

The Centers for Disease Control and Prevention (CDC) provide an overview of the potential harmful biological agents a terrorist organization could acquire, develop, and use against a mass transit facility or network. CDC divides biological agents into categories based on their threat level. Briefly, category A and category B agents are those that can produce the greatest loss of life, severely hamper transportation, and impact the local, regional, and potentially national economy.

Category A agents include organisms that are easy to disseminate, possess high mortality rates, present public health events, create social disruption, and require emergency response and management of event. These agents include *Bacillus anthracis* (anthrax), *Clostridium botulinus* toxin (botulism), *Yersinia pestis* (plague), *Francisella tularensis* (tularemia), and *Variola major* (smallpox).[6] These agents receive the most attention from policymakers and security experts due to their deadly characteristics.

The next category of potential terrorist agents, category B agents, include organisms that are as easy to disseminate as those in category A but result in moderate infection rates and lower mortality rates. This category includes *Brucella* species (Brucellosis), *Salmonella* species, *E. coli, Coxiella burnetii* (Q fever), and Ricin toxin.

Yersinia pestis, Variola major, and *Bacillus anthracis*, if released, will cripple a transportation system and large surrounding areas. *Yersinia pestis* serves as the causative agent for plague, and it is a naturally occurring bacterium that is primary found in wild rodents. Fleas on these rodents serve as the primary vectors of *Y. pestis*, and flea bites transmit this bacterium.[7] Plague manifests in three forms, pneumonic, bubonic, and septicemic, and it is responsible for three pandemics in the modern era, in the mid-6th century, the 14th century (the Black Death), and the early 20th century. Pneumonic plague can spread from person to person via respiratory droplets but usually requires close contact to transmit. However, neither bubonic nor septicemic plague is known to spread from person to person. This is a very lethal disease, sustaining an approximate 50 to 60 percent rate of infection; untreated primary septicemic plague and pneumonic plague remain almost always fatal to infected humans.

Variola virus causes the highly contagious and deadly disease known as smallpox. It infected nearly 300 million people during the 20th century, and scientists and medical personnel believe that it may be the only human infectious disease that has been successfully eradicated.[8] According to the World Health Organization (WHO), the last reported case of naturally occurring smallpox occurred in 1977 in the African country of Ethiopia. *Variola* is also a very lethal disease, and, according to Heymann, *Variola minor* is fatal approximately 1 percent of the time, and *Variola major* is fatal 20 to 50 percent of the time.[9]

Each of these two diseases is very lethal, but there are some key differences in the methods of control for each. First of all, a single case of *Variola* would constitute an emergency and require the implementation of the federal and state smallpox response plans for both surveillance and containment. Also, public health personnel would immediately isolate any individual diagnosed with or potentially infected with smallpox and vaccinate anyone who experienced close contact with the victim. As soon as an outbreak occurred, the federal government would release the U.S. Strategic National Stockpile (SNS) to vaccinate every person in the U.S.[9]

The plague occurs in the U.S. naturally, and an outbreak in the western states among a rodent population would not trigger major concerns. However, if a human outbreak occurred in an area not associated with naturally occurring bouts of *Yersinia pestis*, the CDC recommends isolation of individuals suspected of being infected. In addition, persons who have had close contact with suspected or infected victims would remain under surveillance for the maximum incubation period. Both diseases would place enormous strains on medical and emergency response personnel during disasters, and each requires international notification of a class I outbreak. The plague outbreak would require in-place isolations of affected areas, and a *Variola* outbreak requires immediate isolation of all individuals in the potentially infected environments. In addition, *Variola* outbreaks may require the closing of all transportation routes into and out of potential areas.[10]

The final disease, anthrax, is not a communicable disease, but the CDC still classifies the bacterium as extremely dangerous and continues to list this naturally occurring spore as type A. *Bacillus anthracis* is a gram-positive bacterium, and when individual spores secrete enough of a special toxin, the host develops gastrointestinal, pulmonary (inhalation), or cutaneous anthrax. Anthrax is not contagious, and infection can only occur after coming into contact with a sufficient number of spores. However, it is important to note that spores can live decades in nature or in a laboratory, and micrometer-sized spores can move with clothing, wind, shoes, or on light rail cars to other areas. The *Bacillus anthracis* aerobic bacterium is extremely small, ranging between 1 to 9 micrometers in size.

Bacillus anthracis bacterium can cause gastrointestinal, inhalation, or cutaneous anthrax, but the potential host must ingest, inhale, or physically touch the spores to become infected. The estimated anthrax death rate depends upon the type of *Bacillus anthracis,* the bacteria strain, the diagnosed timeline, and the general health of the victim. Estimates for cutaneous death is up to 20 percent without treatment and less than 1 percent with adequate antibiotics.[11] The gastrointestinal death rate is between 25 and 65 percent. The webpage reference did not provide a specific death rate for inhalation anthrax. Scientific studies to date have not determined how many spores it takes to infect a host with inhalation anthrax, but estimates range from 2,000 to 10,000-plus spores, and lethal infection usually occurs with the inhalation of 12,000 to 20,000 spores.[11] This next section will look at the most reliable existing biological agent detection technology and the path the DHS is taking to pursue a faster and reliable system.

There are additional agents that a terrorist could use in a mass transit attack: cholera, Diphtheria, tuberculosis, yellow fever, viral hemorrhagic fevers (Lassa, Marburg, Ebola, Crimean-Congo, South American, and others not isolated or identified), and severe acute respiratory syndrome (SARS).[6]

Existing Biological Agent Detection Technology

Since the anthrax attacks in 2001, the federal government realized the need to enhance its biological agent detection system, especially in the nation's mass transit facilities. The existing system of healthcare, agricultural and veterinary surveillance and reporting did not provide the needed lead time to execute proper response plans to a clandestine biological attack. Due to the gap in countermeasures and under the auspices of the Office of the Vice President of the United States, the newly formed Department of Homeland Security (DHS) decided to deploy sample units to collect particulates over a specified period and analyze them in a certified laboratory.[12] The White House directed the Department of Defense (DoD) to coordinate with selected subject matter experts (SMEs) and U.S. national laboratories to rapidly deploy a first generation (gen-1) biological detection network. The SMEs deployed the gen-1 system, or biological aerosol sentry and information system (BASIS), to counter the biological threat in high population cities and in mass transit facilities.[13]

Unfortunately, the BASIS units experienced technical and mechanical problems and consequently were not a sustainable system for biological agent detection. Therefore, the federal government searched for a more reliable replacement system that also allowed for polymerase chain reaction (PCR) or other microbiological confirmatory techniques to verify the presence of a biological agent. As a result, the dry filter unit (DFU) was deployed throughout the nation's mass transit facilities.[15] The U.S. military uses the DFU on installations and in other areas to protect troops at home and abroad. Unfortunately, the DFU pulls approximately 1,000 liters of air per minute, which caused concerns in facilities, and the 47-pound units proved difficult to secure during the collection period. In addition, the filters remained too exposed, and the potential for tampering pushed the newly formed BioWatch Program in the DHS to seek a better system.

These concerns about cost-effectiveness drove the SMEs and the accompanying national labs to develop a more reliable unit suitable for indoor use.[16] Presently, the most reliable biosurveillance equipment used today is a filter-based portable sample unit (PSU). HI-Q Environmental Products Company manufactures an all-season-tested PSU device. This popular, outdoor-rated PSU provides continuous air sampling through a ¼-amp or ¾-amp motor operating a vacuum pump system that pulls in up to 100 liters of air per minute through a removable filter.[17] This waterproof indoor/outdoor aluminum-enclosed unit provides users with a small, mobile, and very efficient continually running system. However, the exposed filters require timely collections and replacement, and laboratory personnel must analyze the filters to accurately measure any detectible biological proteins collected during the sampling period. Presently, the DHS BioWatch Program uses these PSUs throughout its national bio-surveillance system. The main disadvantage is the time delay between possible release of a biological agent and the analysis and confirmatory procedures.[18]

The time-delay problem became the focus of many SMEs, and DHS began to seek a replacement for the second-generation (gen-2) biological detection system deployed. One expert provided vital testimony to Congress that outlined the difficulty associated with the present filter collection, analysis, and extended time between potential release of a biological agent and confirmation of that agent's presence in a major transportation facility or city. As a result, the U.S. House of Representatives Subcommittee on Homeland Security received testimony from Dr. Bernard D. Goldstein, Professor of Environmental and Occupational Health at the University of Pittsburgh, on evaluating systems for the early detection of biological threats on February 25, 2010.[14] Dr. Goldstein provided comments on the existing and future biological surveillance systems in use by DHS today. With regard to the timing of the existing biosurveillance systems, Dr. Goldstein said, "With this sampler technology and deployment (known as gen-2), as much as 36 hours may elapse between the collection of genetic material of interest and the availability of essential laboratory test results showing its presence." This time delay is the single driver for the development of a third-generation (gen-3) system that will reduce the testing and analysis time to provide a faster evaluation and decrease the time to treat potentially exposed personnel.

Path to Acquire Autonomous Detection Technology Equipment

DHS attempted to replace the current biodetection system with an autonomous biodetector to reduce the notification timeline, improve the time resolution of the network, and enhance the percentage of population covered. DHS's request for proposals for the gen-3 system calls for a fully autonomous biodetection network that will constantly monitor air in designated cities, mass transit facilities, and other locations throughout the U.S. The operational requirements document (ORD) calls for this system to "rapidly process and accurately analyze aerosol samples, continuously provide reports on system status and processed results, and remain suitable for both indoor and outdoor deployments."[19]

In early 2011, DHS conducted the first part of the acquisition process by completing the phase one field test to determine if technology existed that will support basic continual operations. This test analyzed the baseline assay capabilities of processed vendors and also tested the reliability and maintainability of potential systems. The baseline assay capability determined if the vendors' proposed systems met the required agent detection and verification capability that is so very important to DHS. This extensive evaluation of the assay capability of the potential vendors included "in-chamber" modified agent testing to ensure that sensitivity meets the baseline threshold required in the ORD. Additional testing ensured that the existing detection technology supports the move to an autonomous detection network nationwide.

After this in-depth and complete phase one assay validation and basic reliability and maintainability testing, one vendor received the minimum approval to complete the phase one field test. However, DHS is planning to hold another open process to determine if one or more potential vendors will attempt to satisfy the minimum requirements to compete for and complete the phase two process. The phase two process will include a full operational test and evaluation test plan (OT&E) to determine if units can operate in real time and meet all the operational requirements set forth in the gen-3 ORD.

Presently, DHS is preparing for the investment review board that will approve or deny the application to proceed with the acquisition decision event 2B (phase two process). This decision will allow the DHS to proceed with the process. If the phase two process results in the acquisition of a working unit, higher-risk jurisdictions, mass transit facilities, and other susceptible venues could receive this gen-3 biodetection system. This system could provide mass transit facilities with accurate results as often as every four to six hours, which will result in a faster execution of response and recovery efforts in jurisdictions, states, and the federal government.

CONCLUSION

Biosecurity in multimodal transportation networks requires careful monitoring of high threat areas and coordination with local, state, and federal agencies to operate an early warning system to ensure that all stakeholders can respond to and recover from a biological attack. The general overview of mass transit agencies and their security needs and of the local, state, and federal government's efforts to enhance biosecurity guidance provide a baseline of information needed to continue to research and address biosecurity needs in mass transit networks. The review of the existing biological detection systems and Department of Homeland Security's intentions to move forward (potentially) with the acquisition of the gen-3 system call for continued discussions and decisions to improve mass transit capabilities to identify, respond, and recover from a biological attack. However, additional research and testing are still needed to provide pertinent decision makers with the most reliable and cost-effective biological agent detection system available.

References

1. Jenkins, Brian M., and Gersten, Larry N. Protecting Public Surface Transportation Against Terrorism and Serious Crime: Continuing Research on Best Security Practices. San Jose, CA: Mineta Transportation Institute, San Jose State University (September 2001).
2. Lipton, Eric, and Johnson, Kirk, A Nation Challenged: The Anthrax Trail; Tracking Bioterror's Tangled Course, The New York Times. Accessed at: http://www.nytimes.com/2001/12/26/us/a-nation-challenged-the-anthrax-trail-tracking-bioterror-s-tangled-course.html?fta=y&archive:article_related=&pagewanted=1.
3. U.S. House of Representatives. The Committee on Homeland Security. Detour Ahead, Critical Vulnerabilities in America's Rail and Mass Transit Security Programs. 109th Congress. 2nd Session, (July 7, 2006).
4. Federal Transit Administration, 2006 Transit Watch Toolkit (undated web page). Accessed at: http://transit-fafety-volpe.dot.gov/secruity/Transit Watch/toolkit2006.asp.
5. University of Minnesota Center for Infectious Disease Research and Policy (CIDRAP). Conducted work place discussions (2008–2012). Retrieved from: http://www.cidrap.umn.edu/cidrap/content/bt/bioprep/readings/.
6. Bioterrorism Agents/Diseases, Centers for Disease Control and Prevention (2012). Retrieved from: http://www.bt.cdc.gov/agent/agentlist-category.asp.
7. Yersinia pestis (plague). Center for BioSecurity of UPMC, Biological Threat & Epidemics, Agent Fact Sheets (2012). Accessed at: http://www.upmc-biosecurity.org/website/our_work/biological-threats-and-epidemics/fact_sheets/plague.html.
8. Variola virus (smallpox). Center for BioSecurity of UPMC, Biological Threats and Epidemics, Agent Fact Sheets. Accessed at: http://www.upmc-biosecurity.org/website/our_work/biological-threats-and-epidemics/fact_sheets/smallpox.html.
9. Heymann, D., Control of Communicable Diseases Manual. American Public Health Association. Washington, D.C. (2008)

10. Fact Sheet on Legal Authorities for Isolation/Quarantine. Centers for Disease Control and Prevention (March 9, 2012). Accessed at: http://www.cdc.gov/sars/legal/fs-legal.pdf.
11. Environmental Stability of Diverse Pathogens, Catherine A. Cleland, PhD, Sheila Van Cuyk, PhD, and Kristin M. Omberg, PhD, 2006 Los Alamos National Security, LLC.
12. Fennelly, Kevin P., Davidow, Amy L., Miller, Shelly L., Connell, Nancy, and Ellner, Jerrold J. Airborne Infection with Bacillus anthracis—from Mills to Mail. CDC Emerging Infectious Diseases Online. Accessed at: http://www.cdc.gov/ncidod/eid/vol10no6/pdfs/02-0738.pdf.
13. Wampler, Stephen P. Lab's work provides BASIS for biodetection. Lawrence Livermore National Laboratory (September 7, 2011). Accessed at: https://www.llnl.gov/news/aroundthelab/2011/Sep/ATL-090811_basis.html.
14. Goldstein, Bernard D. Testimony to the Homeland Subcommittee, Committee on Appropriations, U.S. House of Representatives. BioWatch and Public Health Surveillance: Evaluating Systems for the Early Detection of Biological Threats (March 2012). Accessed at: http://www.nationalacademies.org./ocga/testimony/BioWatch_and_Public_Health_Surveillance.asp.
15. Dry Filter Unit. DFU Fact Sheet. Chemical, Biological, Radiological, and Nuclear Information Resource Center (June 2, 2010). Accessed at: https://jacks.jpeocbd.army.mil/jacks/Public/FactSheetProvider.ashx?productId=376.
16. Meehan, Patrick J., Rosenstein, Nancy E., Gillen, Matthew, Meyer, Richard F., Kiefer, Max J., Deitchman, Scott, Besser, Richard E., Ehrenberg, Richard E., Edwards, Kathleen M., and Martinez, Kenneth F. Responding to Detection of Aerosolized Bacillus anthracis by Autonomous Detection Systems in the Workplace. CDC Morbidity and Mortality Weekly Report (June 4, 2004). Retrieved from: http://www.cdc.gov/mmwr/preview/mmwrhtml/rr5307a1.htm.
17. HI-Q Environmental Products Company. PSU-Series: Outdoor, Mobile, Continuous Duty Air Samplers. Accessed at: http://www.environmental-expert.com/STSE_resulteach_product.aspx?cid=3569&idprofile=2620&idproduct=28408&lr=1&idCategory=690.
18. Skinner, Richard L. Inspector General, DHS' Management of BioWatch Program. The Office of the Inspector General, Department of Homeland Security (January 11, 2007). Accessed at: http://www.dhs.gov/xoig/assets/mgmtrpts/OIG_07-22_Jan07.pdf.
19. BioWatch Gen-3 Phase II: Pre-solicitation and Draft Request for Proposal (August 26, 2011). FedBizOpps.gov, Department of Homeland Security. Retrieved from: https://www.fbo.gov/index?s=opportunity&mode=form&id=405828e0ec530c1445821228dbdb54a7&tab=core&_cview=1.

CHAPTER 10

Utilizing Biosecurity Principles to Combat Naturally Occurring Epidemics

J. Craig Reed, Robert A. Heckert, Patricia Delarosa, and Vera Ettenger

INTRODUCTION

Our arsenal against infectious disease is shrinking just as the incidence of emerging disease is rising. We now understand that reactive tools, such as antibiotics and pesticides, act to promote the evolution of more resistant microbes and vectors of disease.[1] Today, we find ourselves at a turning point in the "war" on disease where we need to develop new proactive strategies that address and prevent evolutionary resistance. Biosafety and biosecurity (BS&S) are disciplines focused on preventive strategies that intersect both the public health and biodefense arenas. Biosecurity has long played a role in the prevention and control of infectious disease in agriculture. The World Health Organization (WHO) Food and Agriculture Organization (FAO) biosecurity objectives focus on bioexclusion (the elimination of disease) and biocontainment (the prevention of disease transmission) from a colony, herd, or flock.[2] Agricultural biosecurity decisions are based on the FAO risk analysis for agricultural diseases, a framework for the identification and prioritization of risks and the communication of a science-based process for risk mitigation or elimination (Chapter 8). In this sense, biosecurity defines the prevention and protection strategies employed against microbial threats to plant and animal health, the food supply, or the farm environment.[2]

To date, there is no corresponding public health equivalent to agricultural biosecurity. Perhaps this is because the risk mitigation strategies implemented against infectious diseases on the farm do not directly translate to public health protection (Table 10.1). This gap in BS&S guidance in the area of public health results in biosecurity implementation strategies that do not account for the unique security aspects associated with a living organism, because biosecurity programs in the public health arena are not focused on the prevention of environmental disease circulation. The application of BS&S to public health in a manner similar to the FAO model would drive the integration and communication of infectious disease prevention strategies. It would also add biosecurity to the list of biodefense functions.[2]

SEPARATING DELIBERATE OUTBREAKS FROM NATURALLY OCCURRING EPIDEMICS

Traditionally, the focus of biosecurity has been to separate naturally occurring epidemics from deliberate outbreaks. In this effort, epidemiological investigations carried out by the public health system serve as the foundation of the investigative effort. Deliberate outbreaks have unique implications and profound influences. They cause panic and generate fear and uncertainty. There are several epidemiological indicators that distinguish natural infectious disease outbreaks from the deliberate dissemination of biological agents.[3] Some indicators are highly suggestive that an outbreak is deliberate and requires

TABLE 10.1 Definitions of Biosafety and Biosecurity in Agriculture, Public Health, and Ecology

Sector	Definition
Agricultural	FAO Food Safety Risk Management[2] – Focus on bioexclusion or the elimination of disease and biocontainment or the prevention of transmission of disease from a colony, herd or flock.
Public Health	• BMBL 5th Ed.[3] Biosafety addresses the safe handling and containment of infectious microorganisms and hazardous biological materials for protection of laboratory workers, the community and the environment. • WHO Laboratory Biosafety Manual 3rd Ed.[4] Addresses biology safety and security in the laboratory. • WHO Biorisk Management[5] – The protection, control and accountability for valuable biological materials (VBM) within laboratories, in order to prevent their unauthorized access, loss, theft, misuse, diversion or intentional release.
Environmental	Cartagena Protocol on Biosafety[6] – Need to protect human health and the environment from the possible adverse affects of the products of biotechnology.

additional investigation. Examples include: a disease outbreak uncommon for a certain area; an unusual season or geographical location for the disease; a missing vector in the case of a zoonotic disease that requires one; a cluster of cases with unusual indicators, such as an uncommon causative agent; an unusual disease manifestation (such as inhalational route of infection); an unexplained abundance of dead animals; a large number of casualties; or an unusual disease presentation.[4,5,6] These indicators cannot be viewed as direct evidence of intentional release; they merely suggest that the observed cases are unusual from the public health perspective and require further investigation through a comprehensive epidemiological investigation that will culminate in the development of prevention and mitigation strategies.

There are often similarities between naturally occurring and deliberate outbreaks that can allow for intentional releases to be attributed to natural causes. In 1984 Rajneesh followers deliberately contaminated salad bars at multiple restaurants in Oregon, resulting in more than 750 confirmed case of salmonellosis.[7] At the beginning of the outbreak, public health authorities did not consider that these food-borne salmonellosis cases were the result of a deliberate event, since a large number of *Salmonella* infections occur in the United States annually. The source of the outbreak was puzzling to public health officials because of inconsistencies as compared to other previous natural infections revealed by the epidemiological investigation. The investigation revealed some unusual findings that pointed to a deliberate malicious act: (1) multiple food items were contaminated with the same *Salmonella* strain and (2) restaurant personnel became ill along with the customers. Investigations by the local health department and the USDA revealed that none of the affected restaurants used the same food suppliers and no other restaurants serviced by these same suppliers had contaminated food items. A criminal investigation was initiated due to the size and questionable cause of the outbreak. The fact that the infections were caused by a deliberate malicious act was uncovered because a variety of experts from different agencies were able to collaborate and communicate to trace the source of the organism back of the group responsible for the crime.

THE NEED FOR BIOSAFETY AND BIOSECURITY IN THE RESPONSE TO A NATURALLY OCCURRING OUTBREAK

Just as public health supports biosecurity, so biosecurity can contribute to public health efforts. Select agents (SA) are found in every corner of the world, and the basic technology needed to reproduce these organisms can be purchased in the local hardware or kitchen supply store.[8] The presence of these organisms in the environment contrasts with the strict security and access controls in place in laboratories that store these same organisms. The restricted focus of biosecurity to the laboratory results in a security approach that fails to address the fundamental difference between securing an inanimate object versus a living organism.[9] Although the greatest biosecurity risks, in terms of biological weapon production, are associated with purified laboratory pathogens, vulnerabilities in the public health system also pose significant risks to the development and implementation of medical countermeasures.[10,11] Addressing the "bio" in biosecurity requires that we secure microorganisms that exist and replicate naturally in the environment.

Adding to the problem of microorganism security is the dual-use nature of infectious disease research. Often, scientific equipment and knowledge can be used for good or malicious intent.[12] Our public health security depends on scientific advances in the areas of infectious disease research and medical countermeasures, so it is imperative that a balance be struck between the restrictions and costs imposed by implementation of a biosecurity system and the need to conduct research and publish the results.[13] To protect scientific research while maintaining public security, artificial boundaries, such as the select agent and toxin list, are placed on the regulation of potential bioweapons.[14,15] At the same time, more people from different educational backgrounds are working with infectious organisms or valuable biological materials that could be used to create bioweapons.[16,17] The complex fields of public health and vaccine research, traditionally defined by the core disciplines of bacteriology, virology, parasitology, and epidemiology, has expanded within the past 40 years to include the disciplines of genetics, cell biology, immunology, and synthetic biology.[18] Public health researchers can be found in non-traditional settings such as biotechnology, hospitals, agriculture, and food production as well as in the academic fields of ecology, biochemistry, and bioengineering. In biosecurity, risk is a function of threat, vulnerability, and consequence.[19] In this context, threat is defined as a person or group with capability and intent to do harm; vulnerability is a weakness in the security system that can be exploited; and consequence, when referring to a pathogen, is a function of infectivity, virulence, and availability of effective countermeasures. The increasing numbers of people possessing both technical capability and physical access to infectious organisms or valuable biological materials, coupled with the inability of the regulations to cover all potential threats and vulnerabilities, increases the potential for misuse of a biological agent. Nontraditional roles and settings, not covered by regulations, also raise the possibility that we will not recognize high-consequence risks associated with nontraditional threats or nontraditional laboratories. These problems suggest a need for a change in the way that biosecurity programs are developed and used. One possible change is the application of biosecurity to public health.

Biosecurity is a discipline focused on preventing loss, theft, or diversion of biological materials. In this sense, concepts of biosecurity can be applied to public health to prevent the transmission of disease and, as described in this chapter, may also be applied to preventing the acquisition of increased virulence by an organism.[20] In public health, an effective response to an infectious disease outbreak is focused on disruption of the process of pathogen transmission, as depicted by the chain of infection (Figure 10.1). This chain is composed of links that represent the pathogen (agent), pathogen reservoir (source), portal of exit, transmission, portal of entry, and susceptible host. In an epidemic, the significance of initial infections is often overlooked, or these infections go undiagnosed until the disease becomes established within

FIGURE 10.1 The chain of infection

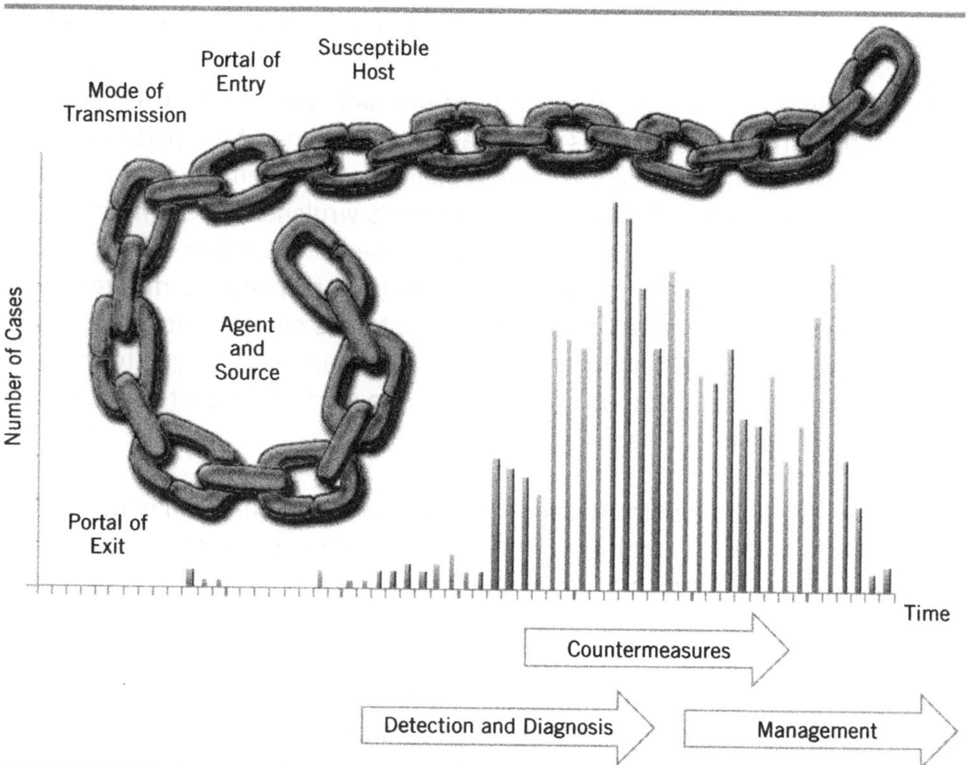

a population. A coordinated response to an epidemic depends on disease detection. Biodefense efforts such as communication, medical countermeasures, and medical control are initiated after the detection of disease within the population. It is possible, however, to break the chain at any link prior to detection and stop the cycle of disease, preventing the chance of an epidemic. To date, biosecurity has not been widely applied to the prevention of disease. In this chapter, methods to expand the role of biosecurity to public health and the prevention of disease will be proposed.

ASSESSING BIOSAFETY AND BIOSECURITY RISKS IN PUBLIC HEALTH: THE SCIENCE OF BIOSECURITY

To understand how biosecurity can be leveraged to support public health efforts, it is necessary to understand how humans co-evolve with their environment both on an ecological and a microbiological level. The fundamental change that is suggested by this exercise is that bio-specific, "micro-level" security controls can be developed based on our understanding of the microbe and the ecology of disease.

Infection and pathogenicity, or the development of disease, are individual responses controlled by diverse factors that include the health and genetic makeup of an individual and the nature of the pathogen.[21] Because there are so many potential causal factors of disease, the reasons why a particular person succumbs to infection while others remain healthy often is, in large part, a mystery. The connection between the pathogen's virulence, or ability to infect the host, and pathogenicity, or the severity of disease caused by the organism, are not straightforward, as every person's immune response is different and every exposure is qualitatively different.[22,34,35] Studies on host-microbe interactions are beginning to elucidate mechanisms that can be leveraged in the prevention of infection and spread of disease. These studies suggest collaborative venues for scientists and security personnel.

In the field of public health it is widely recognized that human health is intricately linked to the environment. Approximately 60 percent of emerging diseases are zoonotic, passing from an animal reservoir to humans.[23,24] Of these, over 70 percent are attributed to direct human contact with wildlife. With the exception of Variola virus (causative agent of smallpox) and rinderpest, natural outbreaks of all select agent organisms occur in nature. Approximately half of all select agent pathogens are zoonotic diseases, including the etiologic agents of anthrax, plague, tularensis, and brucellosis. Depending upon the geographic region of interest, these organisms are often endemic in the wildlife population and frequently pass between wild and domestic animals, spilling over into the human population when conditions allow.[25]

Habitat overlap with a naturally circulating pathogen is associated with an increased incidence of human disease. Humans often enter new environments transiently, exposing themselves to novel pathogens in the process. For example, tourist incursion into Python Cave in Maramagambo Forest, Queen Elizabeth Park, Uganda, has resulted in two reported cases of hemorrhagic Marburg fever including one fatality.[26] These cases are linked to the resident bat population thought to be the source of the infections.

More often, humans engineer an environment that inadvertently places people in contact with pathogens. Increases in Lyme disease cases in the U.S. have been attributed to placement of suburban housing developments near forested areas. The lack of natural predators and hunting in populated areas not only encourages acclimation of deer to humans but also the expansion of the deer population, and their parasites, in areas where humans live. This ultimately results in increased Lyme disease associated with suburban woodlands.[27]

The above examples demonstrate the connection between disease and the environment. In these cases, the fundamental interaction with the pathogen is unchanged; however, technological and engineering advances can also impact the micro-environment of host-pathogen interactions, resulting in microbial resistance to antibiotics or the acquisition of genetic virulence factors via passage between microbes.[28]

In most cases of infectious disease, the greater the pathogen load, the greater the cell damage and the sicker the patient. Disease symptoms are the outward manifestations of the pathogen's damage to the host's cells. In the laboratory, virulence is associated with the ability of the organism to cause laboratory-acquired infections (LAIs). Laboratory procedures that produce aerosols or those that include the use of "sharps" such as needles have been documented to cause LAIs and are known risk factors in the laboratory.[29,30] These risks are highly simplified when applied to public health, however, because they do not account for changes in virulence related to the interaction of the pathogen and the host or the pathogen with the environment. Laboratory strains are not dependent upon transmission from host to host for replication, and so there are differences between transmission of disease in the laboratory and natural disease spread in an epidemic. These differences affect pathogen virulence and, therefore, present different risks.

In nature, a pathogen that can't survive outside of its host must be transmitted to a new host before its current host develops immunity or dies. Pathogens that are completely dependent on their host for survival have evolved to become less pathogenic over time, effectively increasing transmission of the virus by increasing the mobility of the host.[31] Rhinoviruses (causative agents of the common cold) are an example of a pathogen that has traded pathogenicity for mobility. People who are sick with a cold will still go about their daily routines of work and school, exposing many potential hosts to the virus in the process. If a patient becomes immobilized, the pathogen must be spread in a manner that is independent of the host.

There is mounting data to indicate that when a pathogen's survival is not connected to the host's ability to move, the pathogen will become more virulent.[20,31] An example of this is *Vibrio cholera* (causative agent of cholera, an infection that completely incapacitates the host but survives and spreads in polluted water). This model of microbial evolution of virulence has broad applications in biosecurity, as it provides a framework for stratification of pathogens into high-consequence groups by virulence and evaluation of risks associated with working procedures that may result in hospital-associated spread of disease by attending personnel.[32] As a tool in biosecurity, analysis of the pathogen's potential for increased virulence can be used to create a hierarchy of risk in order to focus medical control measures in an outbreak response.

INCREASING SECURITY FUNCTIONS THROUGH THE APPLICATION OF BIOSECURITY TO PUBLIC HEALTH

The history of humans has been shaped by disease. Within the last century, great efforts in disease control have been made. Antibiotics cheaply and easily cure most bacterial infections. Vaccines prevent childhood diseases. Insecticides kill disease-causing insects. With the exception of smallpox eradication by human vaccination, however, none of our medical advances against disease have been permanent. Even the great human achievement of smallpox eradication now appears threatened by a 20-fold increase in the incidence of human monkeypox, a genetically related virus.[33] The growing challenge of emerging and reemerging infectious disease, coupled with a growing number of drug-resistant pathogens, poses challenges for both the public health and biodefense communities. Biosecurity on the farm has been used successfully in the prevention of disease. This approach adds a layer of security extending out from the veterinary laboratory to the agricultural community.

Application of biosecurity measures to public health can be accomplished, creating security functions that target macro- and micro-drivers of disease, increasing public health security in the community and providing bio-specific prevention measures to the biosecurity toolkit. Another benefit of a biosecurity/public health collaboration would be to leverage cross-disciplinary communication in the life sciences using science-specific security measures with the goal of building bridges between the science community, policymakers, first responders, and law enforcement. These macro- and micro-security measures will come from the research in host-pathogen interactions, providing dynamic, science-based solutions for disease prevention.

CHALLENGES ASSOCIATED WITH THE INTERNATIONAL COORDINATION OF OUTBREAK RESPONSE

During a natural outbreak of disease in a human or animal population there could potentially be many different challenges associated with international coordination of biosecurity of the reservoir, the site, or the samples. These challenges range from conflicting jurisdictional issues to ownership of organisms and data. Although it may seem apparent that biosecurity coordination and collaboration should be paramount during an outbreak, in reality this will be very problematic to accomplish. Some of the issues that must be addressed within and among governments include:

1. If there is to be any coordinated effort, the countries involved must have an internal command and control structure with a specific agency possessing dedicated responsibility for domestic biosecurity functions. Because diseases spread so rapidly with no respect for borders, there must also be an agency dedicated to the coordination of the country's international biosecurity efforts. This may or may not be the same agency that has responsibility for domestic biosecurity functions. The World Health Organization (WHO) may also play a role in the coordination of biosecurity of the site, samples, and reservoirs (people and animals). However, countries deal with disease control in many different ways, and most may wish to implement their own biosecurity control strategies rather than have them dictated or directed by the WHO. Indeed, regions within each country may make a case for local control over their local situation. Without effective federal-level involvement this will lead to significant erosion of coordination and, potentially, the loss of biosecurity altogether.

2. If the outbreak is thought to be terrorism, other international or domestic security agencies may wish to control the biosecurity of the site, country, or region. These agencies will see biosecurity very differently than a health agency. Intelligence and security agencies will be concerned with the process of evidence gathering, chain of custody, attribution, etc., and not at all concerned with the transmission mechanisms of the disease. The activities and coordination efforts of security agencies will be very different than that of the WHO or a domestic health agency. In fact, the movement of security personnel, the processes of sample collection and disposition, and restrictions on the involvement of scientific personnel may not only actually hamper disease control efforts but may also lead to further outbreaks.
3. No biosecurity plan or process for an outbreak has ever been written or agreed upon. This however, does not mean that there aren't several different visions of what the process and outcomes of international biosecurity coordination should be. During an outbreak a variety of domestic and international organizations may try to impose their version of biosecurity on the situation. This then requires the coordinator (whomever it may be) to pick from one of the strategies being imposed, choose a hybrid strategy, or develop a new plan for security of the site, the samples, and the people. This lack of international consensus on the biosecurity process and its outcomes is the biggest international challenge to achieving a coordinated international biosecurity response for an international outbreak of disease. Perhaps the World Health Organization (WHO) can facilitate an international effort on this front.
4. If the outbreak has spilled beyond the border of any one country, the coordination requires all of the above aspects to be fulfilled and excellent systems by which to implement all aspects of the biosecurity plan to be developed. An effective biosecurity plan for international response must involve background checks of individuals involved in sample collection, transport, and patient/animal security as well as agreed-upon protocols and standard operating procedures (SOP) for sample collection, transport, testing, and confirmation; physical security of patients, animals, and/or site; and management, storage, and sharing of data and biological materials.
5. Because there is no agreed-upon process or definition of biosecurity during an epidemic, there is also no agreed-upon list of organisms that necessitate a biosecurity response. Further, there is no agreement on the definition of or need for a gradient of biosecurity response or mechanism

by which biosecurity controls are to be coordinated. Every country and region of the world has a different opinion of which pathogens are significant and may require some degree of biosecurity. This is due to differences in endemicity and disease control strategies employed around the world. In some countries many very virulent and significant diseases are virtually nonexistent or at low levels (e.g., human tuberculosis, human immunodeficiency virus, foot and mouth disease, etc.), while in other countries these diseases are very common and exist in the normal population. Therefore, it is difficult if not impossible to establish universally applicable biosecurity measures based solely on the virulence or pathogenicity of an organism. The only two diseases that require immediate biosecurity are smallpox and rinderpest. There should be universal concurrence because the causative agents of these diseases no longer exist in nature.

6. Depending upon the size and severity of an epidemic, many international aid agencies, national authorities, nongovernmental organizations, and global health organizations can be expected to send representatives to different regions or focal points to assist with disease control and data collection efforts. It will become very difficult to coordinate their activities during sample collection, transport, testing, confirmation, and information storage. All will have a legitimate reason to be present to conduct their activities. And some may not recognize any authority's right to impose restrictions or limitations on their activities. Therefore, the level of biosecurity they will practice or agree to abide by will vary based upon their training, background, understanding, and appreciation of the importance of biosecurity.
7. During animal disease outbreaks it's common for owners of small farms to move their livestock in an attempt to hide the disease. This defeats biosecurity processes and permits the disease to spread further. This was evident in British Broadcasting Corporation news reports during the foot and mouth disease outbreak in the United Kingdom in 2001.[36] Livestock often move across national boundaries to areas where disease control policies may be less stringent or nonexistent. Due to "leaky borders" and a lack of biosecurity coordination between countries, "transboundary animal diseases" (as referred to by the Food and Agricultural Organization) tend to flourish during times when no outbreaks are occurring and become explosive during outbreaks. Better coordination among international livestock agencies (e.g., sheep producers, poultry producers, cattle associations, etc.) and better coordination among countries

that share common borders will simultaneously enhance disease control and eradication efforts and biosecurity efforts.

8. With public health epidemics people become the reservoirs of disease. People (sick or healthy) are constantly moving locally, regionally, and nationally, and, in so doing, they take their pathogens with them. Implementation of quarantine measures (to ensure biosecurity) is often difficult if not impossible. During the SARS epidemic of 2003 the SARS virus moved easily from country to country, while control measures failed until more quarantine measures could be implemented. The voluntary quarantine of animals or people has always been difficult and contentious due to the lack of international agreement and cooperation in battling the spread of disease.
9. The creation of bureaucratic processes to coordinate and control human responses to disease outbreaks is necessary, but it is virtually impossible to limit the movement of infected wildlife or insect reservoirs or vectors. During the recent avian influenza global pandemic scare, migratory birds were suggested as a possible carrier for the virus. Fortunately, migratory birds did not turn out to be a large factor in the spread of the disease, but in a future pandemic this may not be the case. In the case of aquatic diseases, how can biosecurity of disease be implemented in the seas?

The challenge of coordinating international efforts to define, implement, and monitor biosecurity is not insignificant. The issues outlined above are very real, and the desire to develop a coordinated biosecurity process remains elusive. All parties involved in international pandemic control must come together before a pandemic outbreak occurs and begin to discuss and achieve consensus on biosecurity coordination. These conversations could be facilitated by the WHO. At a future point, tabletop exercises could help identify major issues and lead to preemptive solutions.

BIOSECURITY IMPEDIMENTS TO INTERNATIONAL HEALTH PROGRAMS

Organizations such as WHO, the Pan American Health Organization, the United Nations Children's Fund, and the U.S. Departments of Defense, State, and Health and Human Services, including the Centers for Disease Control and Prevention and the National Institutes of Health, as well as many nongovernmental organizations (NGOs) and private voluntary organizations (PVOs) have created, funded, implemented, and supported programs to prevent and control infectious diseases for many years. Most international health programs that respond to natural outbreaks rely on the open exchange of information, a large degree of collaboration between multiple donors, large numbers of paid and unpaid personnel, and a huge flow of people, equipment, animals, and samples to and from affected areas. All these elements are traditionally used by international health programs to work with domestic organizations to control an outbreak. Unfortunately, all are generally counter to good biosecurity principles and practices. Therefore, the current approach used to rapidly and effectively monitor, control, and stop a natural disease outbreak is at odds with the biosecurity principles to be applied during an outbreak.

It is critical to human health that established international health programs function during a crisis. However, there are many significant issues that arise regarding biosecurity implementation while establishing effective international health programs. There are few to no international health programs that integrate the entire spectrum of biosecurity including site biosecurity, specimen collection and chain of custody, diagnostic laboratory operations, sample storage procedures, and activities involved in outbreak control. Furthermore, these activities are not fully coordinated with other international health organizations also involved in outbreak response. Therefore, all of the issues described above regarding the challenges of coordinating biosecurity during an outbreak apply to the implementation of international health programs as well.

Good biosecurity and good international health programs are not always in harmony. While certain features of disease control support some aspects of biosecurity, there are many other parts that are in conflict. For example:

1. International health programs want to stop the spread of disease as quickly as possible and therefore support the quarantine of people and animals. However, due to the logistics of disease control, international health programs often increase the movement of people and animals (at least in some local areas). Often healthcare workers arrive from other countries to provide aid and then return to their home country or locale. If this is not done carefully with biosecurity in mind, disease may spread inadvertently. Some countries and some states have developed guidance for some diseases (e.g., Hawaii for avian influenza [http://hawaii.gov/hdoa/ai/WorkerProtection]), but these are haphazard guidance documents, for which there is not always wide awareness, acceptance, or application.
2. International health programs need to monitor disease statistics and track the epidemiological course of the disease. To accomplish this, they

must collect a large amount of data on the location of the pathogen. This is in direct conflict with good information security practices, where the location of highly dangerous organisms is not a subject for public announcement.

3. International health programs increase disease surveillance activities during and after an outbreak. This results in increased traffic to and from towns, villages, and farms. The increased movement of personnel increases the likelihood that disease will spread from an infected locality to an uninfected neighbor. All of this activity provides an opportunity for anyone with malicious intent to gain access to a large amount of the infectious agent. For example, during a foot and mouth disease outbreak, anyone taking a sample from a vesicle of an infected cow would immediately have 100 million virus particles for use. The significant increase in the number of individuals entering the area of the outbreak is at odds with good biosecurity practices.
4. During outbreak monitoring and control there is a huge surge in sample collection, transport, testing, and storage. These activities lead to increased opportunity for samples to be misplaced, stolen, or misused. Biosecurity of the sample is often forgotten in the haste to deal with samples and get test results. This represents a significant vulnerability.
5. During an outbreak a large number of people often present themselves to hospitals for treatment or diagnosis. This surge in human traffic, along with the pathogens they carry, is in direct conflict with prudent biosecurity and can in some cases lead to further disease spread, as was seen in the SARS epidemic.[37] In addition, biosecurity measures at some hospitals or facilities receiving patients with high consequence pathogens are variable.[38]

CONCLUSION

If biosecurity programs are fully implemented during a disease outbreak, spread of the disease is likely to be severely curtailed. For example, offers of aid could be refused during the pandemic in order to establish better biosecurity of the personnel involved or the access to samples; however this many not be in the best interest of the international community, which wants the pandemic stopped as soon as possible. So the question remains of how to get the appropriate amount of international aid and yet maintain an appropriate balance with the requirements of biosecurity. There are both advantages and disadvantages to having large international health programs involved in outbreak control. Nonetheless, the activities need to be carefully implemented to ensure that the outbreak is not compounded through a lack of proper biosecurity.

Increased awareness of biosecurity principles and practices on the part of all personnel representing responding government agencies, international aid agencies, and international health programs will be a significant step in the right direction. The principles and practices of good biosecurity must be implemented at sites of outbreak all the way to diagnostic laboratories and must be built into all phases of the international health agencies' response to an outbreak. These biosecurity measures must be codified into written biosecurity program plans and standard operating procedures to ensure consistency and uniformity of application. The program plans and SOPs must be communicated to all organizations responding to an outbreak, and their staff must receive training so there are no differences in implementation in the field. All of this requires a greater spirit of collaboration and cooperation among governments, aid groups, and international agencies.

References

1. Harrison, Polly F., and Lederberg, Joshua (eds.). National Academy of Sciences. Institute of Medicine, National Academy Press, Washington, D.C. (1998). Accessed at: http://www.nap.edu.
2. World Health Organization. Food and Agriculture Organization of the United Nations Rome. Report of a Joint FAO/WHO Consultation. Risk management and food safety. FAO Corporate Document Repository. M-83 (1997) . Accessed at: www.fao.org/docrep/24982e00.htlm.
3. Carus, W.S., Working Paper: Bioterrorism and Biocrimes. The Illicit Use of Biological Agents Since 1900. Center for Counter proliferation Research, National Defense University. Washington D.C. (2001).
4. Fine, A., and Layton, M. Lessons from the West Nile viral encephalitis outbreak in New York City, 1999: Implication for bioterrorism preparedness. Clin. Infect. Dis. 32: 277–282 (2001).
5. Gurskey, E. Anthrax 2001: Observations on the medical and public health response. Biosecur. and Bioterror. 1(2):97–110 (2003).
6. Crupi, R.S., Asnis, D.S., Lee, C.C., Santucci, T., Marino, M.J., Flanz, B.J. Meeting the challenge of bioterrorism: Lessons learned from West Nile virus and anthrax. Am. J. Emer. Med. 21(1):77–79 (2003).
7. Török, T.J., Tauxe, R.V., Wise, R.P., Livingood, J.R., Sokolow, R., Mauvais, S., Birkness, K.A., Skeels, M.R., Horan, J.M., Foster, L.R. A large community outbreak of Salmonellosis caused by intentional contamination of restaurant salad bars. JAMA 278(5), pp. 389–395 (1997).
8. Rambhia, K.M., Ribner, A.S., Gronvall, G.K. Everywhere you look: select agent pathogens. Biosecur. Bioterror. 9:69–71 (2011).
9. Chyba, C.F. Biological security in a changed world. Science 293(5539):2349 (28 Sept. 2001).

10. Garrett, L. Betrayal of Trust: The collapse of global public health. Chapter One: Filth and Decay. Hyperion. New York (2000).
11. Lawler, A. Boston weighs a ban on biodefense studies. Science 304:665 (2004).
12. Wallerstein, M.B. Scientific Communication and National Security. Science 224:460 (1984).
13. Sta, Ana J.L., Frankel, M.S., and Berger, K.M. Educating scientists about dual use. Science 326:1193 (2009).
14. U.S. Department of Agriculture. Animal and Plant Health Inspection Service. APHIS Select Agents Regulations. Federal Register. 7 CFR Part 331 and 9 CFR Part 121(2008).
15. Department of Health and Human Services. Centers for Disease Control. CDC Select Agents Regulations. Federal Register 42 CFR Part 73 (2008).
16. King, D.A., Peckham, C., Waage, J.K., Brownlie, J., Woolhouse, M.E.J. Infectious Diseases: Preparing for the future. Science 313:1392–3 (2006).
17. Fauci, A.S. 2001. Infectious Diseases: Considerations for the 21st Century. Clin. Infect. Dis. 32:675–85 (2001).
18. Tumpey, T.M., Basler, C.F., Aguilar, P.V., Zeng, H., Solorzano, A., et al. Characterization of the resconstructed 1918 Spanish Influenza Pandemic Virus. Science 310:77–80 (2005).
19. World Health Organization. WHO Biorisk Management: Laboratory biosecurity guidance (2006). Accessed at: http://www.who.int/csr/resources/publications/biosafety/WHO_CDS_EPR_2006_6.pdf.
20. Walther, B.A., Ewald, P.W. Pathogen survival in the external environment and the evolution of virulence. Biol. Rev. 79: 849–69 (2004).
21. Aragon, T., and Reingold, A. University of California, Berkeley. Center for Infectious Diseases and Emergency Readiness. Epidemiologic Concepts for Controlling Microbial Threats (2010). Accessed at: http://www.idready.org/documents/Controlling_Microbial_Threats_booklet.pdf.
22. Cassadevall, A., and Pirofski, L. Host-Pathogen Interactions: The attributes of virulence. J. Infec. Dis. 184:337–44 (2001).
23. Jones, K.E., Patel, N.G., Levy, M.A., Storeygard, A., et al. Global trends in emerging infectious disease. Nature 451:990–3 (2008).
24. Taylor, L.H., Latham, S.M., Mark, E.J. Risk factors for human disease emergence. Phil. Trans. Royal Soc. London. Series B: Biological Sciences 356:983–9 (2001).
25. Daszak, P., Cunningham, A.A., Hyatt, A.D. 2000. Emerging infectious diseases of wildlife-threats to biodiversity and human health. Science 287:443 (2000).
26. Timen, A., Marion, P.G., Koopmans, A., et al. Response to imported case of Marburg Hemorrhagic Fever, the Netherlands. Emer. Infec. Dis. 15(8):1171–75 (2009).
27. LoGiudice, K., Ostfeld, R.S., Schmidt, K.A., and Keesing, F. The ecology of infectious disease: Effects of host diversity and community composition on Lyme disease risk. Proc. Nat. Acad. Sci. 100(2):567-571 (2003).
28. Wilson, B.A., Slayers, A.A., Whitt, D.D., Winkler, M.E. Bacterial Pathogenesis: A molecular approach, 3rd ed. ASM Press (2010). .
29. Sewell, D.L. Laboratory-associated infections and biosafety. Clin. Micro. Reviews, pp. 389–405 (July 1995).
30. Sulkin, S.E., Pike, R.M., Survey of laboratory-acquired infections. Am. J. Pub. Health and the Nation's Health 41:769–81 (1951).

31. Ewald, P.W. 1996. Guarding against the most dangerous emerging pathogens. Emerg. Infect. Dis. 2:245–57 (1996).
32. Girou, E., Legrand, P., Soing-Altrach, S., Lemire, A., et al. Association between hand hygiene compliance and methicillin-resistant Staphylococcus aureus prevalence in a French rehabilitation hospital. Infect. Control and Hosp. Epidem. 27:1128–30 (2006).
33. Rimonin, A.W., Mulembakani, P.M., Johnston, S.C., Lloyd Smith, J.O., et al. Major increase in human monkeypox incidence 30 years after smallpox vaccination campaigns cease in the Democratic Republic of Congo. Proc. Natl. Acad. Sci., USA 107:16262–7 (2010).
34. Watson, D.W. and Brandley, C.A. Virulence and Pathogenicity. Ann. Rev. Microbiol. 3: 195–220 (1949).
35. Casadevall, A., and Pirofski, L.A. Host-Pathogen Interactions: Redefining the basic concepts of virulence and pathogenicity. Inf. Immun. 67(8): 3703–13 (1999).
36. BBC news report. Accessed at: http://news.bbc.co.uk/2/hi/uk_news/2667343.stm.
37. Toner, E.S., and Nuzzo, J.B. Acting on the Lessons of SARS: What Remains To Be Done? Biosecurity and Bioterrorism: Biodefense Strategy, Practice, and Science. June 9(2): 169–174. doi:10.1089/bsp.2010.0074 (2011).
38. Puro, V., Fusco, F.M., Schilling, S., Thomson, G., De Iaco, G., Brouqui, P., Maltezou, H.C., et. al. Biosecurity Measures in 48 Isolation Facilities Managing Highly Infectious Diseases. Biosecurity and Bioterrorism: Biodefense Strategy, Practice, and Science. 10(2): 208-14. doi:10.1089/bsp.2011.0098 (2012).

PART IV

BIOSECURITY POLICY, BIOTERRORISM, AND THE FUTURE

Current Efforts to Fortify Biosecurity Policy in the U.S.

Gigi Kwik Gronvall

INTRODUCTION

U.S. biosecurity policies have expanded over the past decade. This expansion has evolved relatively quickly. When most laboratory leaders and university professors were trained, security concerns about pathogens were nearly absent. Research could be performed on pathogens such as *Bacillus anthracis,* the causative agent of anthrax disease, without federal security clearances. New biosecurity concerns are not only governing what type of and how research is conducted, but changes in security policy are now dictating which research manuscripts are published. Given the rapidly changing landscape of biosecurity policies and regulations, it is critical for scientists and administrators to stay abreast of new developments. These developments affect the practice and funding of infectious disease research.

This chapter describes U.S. efforts to fortify and clarify biosecurity policies in three major areas of biosecurity: (1) safe and secure handling of pathogens that could be used as biological weapons, (2) preventing dual-use research of concern (DURC) from misuse, and (3) oversight of the emerging field of synthetic genomics. The appendix to this chapter also includes a timeline of selected events in biosecurity policy that details its evolution and provides significant reports for further reading (Appendix 2).

EMERGING POLICIES LIMITING ACCESS TO PATHOGENS AND ENHANCING LABORATORY SECURITY

In the U.S., access to pathogens is regulated through the select agent program, which was launched by the Antiterrorism and Effective Death Penalty Act of 1996.[1] The Act prohibited the transfer of some "select" agents (see also Chapters 2 and 7) from one laboratory to another without registration with the Centers for Disease Control and Prevention (CDC). Congress passed this law in response to the difficulties in prosecuting Larry Wayne Harris, a member of the white supremacist group Aryan Nations, who in 1995 legally obtained *Yersinia pestis*, the causative agent of plague, from American Type Culture Collection (ATCC), a biological services company. After the anthrax letter attacks in October 2001, the select agent program was significantly enhanced by the 2001 U.S. Patriot Act[2] and the Public Health Security and Bioterrorism Preparedness and Response Act of 2002.[3] Both Acts strengthened accountability measures and restricted access to agents on the select agent list. The new program was implemented in stages starting in 2003, with final rules for the possession, use, and transfer of select agents and toxins (42 CFR Part 73, 7 CFR Part 331, and 9 CFR Part 121) published in the Federal Register on March 18, 2005.[4]

The select agent program involves three government agencies: the Animal and Plant Health Inspection Service (APHIS) at the Department of Agriculture (USDA), the Centers for Disease Control and Prevention (CDC) at the Department of Health and Human Services (HHS), and the Department of Justice (DOJ). The CDC regulates some biological select agents and toxins (BSAT), such as Ebola virus, while APHIS regulates others, including foot and mouth virus. Some agents, such as anthrax, are "overlap" agents, which can be regulated by either CDC or APHIS. These agencies also inspect facilities conducting research on select agents. In 2008, approximately 12,000 individuals at institutions conducting work on BSATs were enrolled in the select agent program.[5] Currently, there are approximately 80 regulated pathogens and toxins on the select agent list.

The select agent program has made research more tractable and accountable. At the time of the anthrax letter attacks the CDC could not definitively state who had access to laboratory stocks of *Bacillus anthracis*, as there was no accountability system. Now there is a list containing information on who has access to particular pathogens, and it is illegal to acquire BSATs without special clearance. However, in spite of its importance, security efforts are inherently limited;

the select agent program only applies to U.S. laboratories, tracking individual pathogens is extremely difficult, and most of the select agent pathogens occur naturally.[6] With the exception of the smallpox and 1918 influenza viruses, all of the pathogens on the list can be found in nature and routinely cause disease in people or in animals. This has been demonstrated through a scan of disease surveillance reports from such sources as the ProMED Mail archives, the World Animal Health Information Database (WAHID) of the World Organization for Animal Health (OIE), and the Global Early Warning and Response System for Major Animal Diseases including Zoonoses (GLEWS).[7] Within the span of two years, there were reports of natural disease caused by nearly all of the select agents, and it can be assumed that the actual rate of disease incidence is higher than surveillance reports indicate.[7]

Some changes to the select agent program are forthcoming: on June 13, 2011, the Federal Experts Security Advisory Panel (FESAP) released its Recommendations Concerning the Select Agent Program,[8] as called for in Executive Order 13546.[9] The committee recommended that the select agent list be stratified to create tier 1 pathogens to which access would be subject to enhanced scrutiny and recommended that 25 pathogens be removed from the list. Indeed, on October 5, 2012, CDC and APHIS issued the results of the biennial review of the select agents and toxins rules and requirements.[10,11] Certain BSATs previously listed were removed, while new BSATs were added. In addition, new security requirements for tier 1 BSATs were included.

Additional policies may be enacted to prevent insider threats (see Chapter 6). The insider threat came under scrutiny when the Federal Bureau of Investigation (FBI) declared in 2008 that Bruce Ivins, a senior scientist at the United States Army Institute for Infectious Diseases (USAMRIID), was solely responsible for the 2001 anthrax attacks.[12] The insider threat has remained a primary focus of many policy discussions and reports (see Appendix 2).[9,13,14,15,16]

Many laboratory measures set forth by the select agent program are intended to cull potential security risks in the hiring process. Since 2003, a scientist must go through a clearance process to have access to BSATs. This clearance process is roughly equivalent to that required for classified information.[13] Often, research institutions storing BSATs have additional procedures (i.e., credit and reference checks) that go beyond federal requirements.

Nonetheless, it is possible to add other regulations to assist insider threat prevention, including behavioral monitoring and assessment of laboratory workers.

The Federal Experts Scientific Advisory Panel (FESAP) is currently examining whether psychological assessment of personnel working with BSATs should be a national requirement and "will further explore the utility of behavioral assessments to identify indicators of potential for violent behaviors, criminal behaviors, or other behaviors that pose a national security risk."[15] Some institutions have already implemented behavioral monitoring; all NIH researchers who work in maximum containment (BSL-4) undergo annual behavioral health screens "designed to help assess the workers' psychological resilience and individual attitudes towards laboratory safety and personal responsibility."[17]

Nonetheless, requiring behavioral assessments of scientists is controversial.[18,19,20] A National Academies of Science (NAS) committee examined personnel reliability measures, often referred to as "biosurety," (see Chapter 1) and determined that such a system was not needed and may be counterproductive.[14] Moreover, the screening procedure might be unfair or too intrusive and "ironically contribute to someone becoming disgruntled and potentially susceptible to the very behavior screening is intended to prevent."[14]

Enhancing personnel reliability policies could also create a false sense of security within research institutions. Experts in psychology of terrorism "have been nearly unanimous in [the] conclusion that mental illness and abnormality are typically not critical factors in terrorist behavior" and that what is characteristic is a terrorist's *normality* in spite of that individual's performing a crime.[21] The counterproposal to increased behavioral monitoring is to address the insider threat with more active laboratory leadership, trust, and openness—that is, giving laboratory leaders the training, responsibility, and time to take an active role to observe, evaluate, and personally monitor what is happening in their laboratories on a daily basis.[18] Given that research on insider threats suggests that there are usually warning signs prior to an incident and that it is genuinely rare for a spontaneous breach of an insider's trust to occur, a laboratory leader's presence (coupled with training to recognize signs of stress and mental illness) is likely to be far more effective than an annual psychological exam in predicting security threats.[13]

The debate over the extent of security measures and oversight needs to be balanced with the consequences of potentially having too much regulatory policy. Otherwise, the situation might yield counterproductive results for biosecurity. There are direct burdens on laboratories and researchers. For example, multiple, overlapping inspections take weeks to prepare for, some inventory control measures are not biologically relevant since microorganisms multiply, and long waiting periods before scientists are cleared before beginning work decrease productivity. These measures can be counterproductive in the larger picture of vaccine development, treatment, and diagnostic testing against biological organisms of concern. Less obvious are the impacts on biosecurity due to low work performance or the unattractiveness of biodefense and its staunch regulations.[20] Ultimately, according to the Defense Science Board, which examined the issue of laboratory security and monitoring of personnel, "a determined adversary cannot be prevented from obtaining very dangerous biological materials intended for nefarious purposes. We need to recognize this reality and be prepared to mitigate the effects of a biological attack. We, as a nation, are not prepared."[22] The security that can be bought by enhancing the rigor of the select agent program may inadvertently be a detriment to preparedness without any accompanying benefits to security.

POLICIES THAT FOCUS ON DUAL-USE RESEARCH OF CONCERN

Dual-use research has sparked rapid growth and attention in biosecurity policy development. The National Science Advisory Board for Biosecurity (NSABB) defines dual-use research of concern (DURC) as "research that, based on current understanding, can be reasonably anticipated to provide knowledge, products, or technologies that could be directly misapplied by others to pose a threat to public health and safety, agricultural crops and other plants, animals, the environment, or materiel." [23]

Recent policy discussions of DURC in 2011–2012 focused on H5N1 influenza virus (see Chapter 7) after two research groups sought to publish how the H5N1 influenza virus could become more transmissible.[24] In response to the H5N1 influenza studies, the NSABB took the unprecedented step of denying publication but recanted after several months of controversial debate in the global scientific community.[25] Ultimately, the research papers were allowed to be published.[26,27]

The H5N1 controversy was the catalyst for the United States Government Policy for Oversight of Life Sciences Dual Use Research of Concern, released on March 29, 2012.[23] This new policy requires U.S. government funding agencies to review all proposed and funded projects to identify dual-use concerns and propose risk mitigation plans that could include voluntary redaction or classification of research publications.[23] Anthony Fauci, the Director of the National Institute of Allergy and Infectious Diseases at NIH, revealed that the

NIH anticipated new guidelines for research institutions handling dual-use research of concern (DURC), open to public comment.[28] The policy is limited to research involving 15 pathogens already classified as BSATs, which are already under considerable oversight. While the new policy limits the burden on scientific research, it also limits its potential effectiveness.[29]

The dilemma over dual-use research is likely to continue over time, along with a mix of cultural expectations, best practices, as well as unencountered examples of dual-use research. DURC is an extremely difficult area to regulate.[30] In contrast to the "command and control" regulatory regimes governing nuclear weapons research, it is nearly impossible to assign sharp edges to what type of biological research is permissible or impermissible. For most DURC, there are legitimate, beneficent reasons why certain experimentation was pursued and why the general public needs to know results of this research. Given that a great swath of biological research could be deemed DURC, picking the scientific areas for increased regulation that would be useful for a terrorist requires "extraordinary visibility into the actions of those who would misuse biology."[31]

Decisions on how to regulate DURC are likely to remain controversial, influenced by discord on its importance, the likelihood that it could lead to weapons development, and threats of terrorism. They are likely to be highly subjective judgments. The controversy over H5N1 papers within the NSABB, World Health Organization, The Royal Society, and the National Academies of Science[32] speaks to the complexity surrounding dual-use research and the difficulty of reaching any agreement.

EVOLVING OVERSIGHT OF SYNTHETIC BIOLOGY

Security concerns regarding synthetic biology have been present nearly from its inception. In July 2002, researchers from the State University of New York made headlines when they created a polio virus from scratch.[33] The publication caused alarm, especially when it became widely known that the genetic information for polio virus, as with most other viruses including smallpox, is freely available on the Internet. Representative Dave Weldon (R-FL) introduced a resolution criticizing the publishers of Science for releasing the paper and DoD for funding the work and not classifying the report.[34]

The series of potentially problematic papers did not abate. Less than two years after the publication of the poliovirus paper, a research team based at the

J. Craig Venter Institute (Rockville, MD) published its de novo synthesis of the 5,486 bp bacteriophage φX174 (the genome of a simple virus), which took them just two weeks.[35] Polio virus, though larger in size (polio was 7500 bp), required three years of work to yield an infectious virus.

Of course, creating a viable pathogen in the laboratory is not technologically simple, and assembling the genetic material that encodes the pathogen is just an initial step. However, the process of making a pathogen from scratch is difficult, but the tools needed are increasingly accessible and affordable. The growing scientific DIY bio movement has added to policymakers' concerns that soon the ability to make a pathogen could be facilitated by interested people with minimal training.

In response, the first report of the Presidential Commission for the Study of Bioethical Issues addressed synthetic biology.[36] The commission conducted a wide-ranging review of synthetic biology research and practice, including hearings, and issued 18 recommendations, including more coordinated federal oversight of scientists working in research institutions. They took a measured approach and did not do what a number of environmental groups called for—a moratorium on the commercial use and release of synthetic organisms until regulations could be developed and potential safety concerns addressed.[37]

On October 13, 2010, Health and Human Sciences (HHS) released its Screening Framework Guidance for Providers of Synthetic Double-Stranded DNA. The guidance asks suppliers of double-stranded DNA to screen orders against GenBank, the NIH genetic sequence database, which is an annotated collection of all publicly available DNA sequences. Suppliers were instructed to screen customers as well to ensure compliance with U.S. trade restrictions and export controls. If sequence screening determines that the genetic material request is only available to those with select agent clearance, then the customer must be in compliance with select agent regulations.

At this time, stricter regulations would be difficult to put into practice without imposing greater burdens on U.S. scientists doing legitimate research.[39] Gene synthesis is an international business, which means that many companies are not subject to U.S. regulations. If it becomes too onerous to go through commercial suppliers, scientists may make the genes they need themselves, obviating the usefulness of regulations on commercial suppliers.

CONCLUSION

Development of biosecurity policies relating to pathogen access, management of dual-use research, and oversight of emerging fields, such as synthetic genomics, is still in progress and likely to evolve in the coming years. Changes in biosecurity policies have the potential to drastically affect how research is conducted and funding allocated. It is imperative that scientists stay updated on current security legislation.

References

1. The Antiterrorism and Effective Death Penalty Act 110 Stat. 1214 (1996).
2. Uniting and Strengthening America by Providing Appropriate Tools Required to Intercept and Obstruct Terrorism (U.S. Patriot Act) Act of 2001, §Sec. 817. Expansion of the Biological Weapons Statute (2001).
3. Public Health Security and Bioterrorism Preparedness and Response Act (2002).
4. Centers for Disease Control and Prevention. Select Agent Registry. 2012. Available at: http://www.selectagents.gov/AboutUS.html (accessed July 19, 2012).
5. Weyant, R. National Science Advisory Board for Biosecurity meeting (December 10, 2008.) Available at: http://oba.od.nih.gov/biosecurity/meetings/200904/12-10-08%20NSABB%20Minutes_FINAL-complete.pdf (accessed July 19, 2012).
6. Gronvall, G.K. Improving the Select Agent Program. Bulletin of Atomic Scientists Web Edition (October 28, 2008). Available at: http://www.thebulletin.org/web-edition/op-eds/improving-the-select-agent-program (accessed July 19, 2012).
7. Rambhia, K.J., Ribner, A.S., Gronvall, G.K. Everywhere you look: Select Agent pathogens. Biosecur Bioterror. Mar 9(1):69–71.
8. Federal Experts Security Advisory Panel. Recommendations Concerning the Select Agents Program (June 13, 2011). Available at: http://www.phe.gov/Preparedness/legal/boards/fesap/Documents/fesap-recommendations-101102.pdf (accessed May 15, 2012).
9. President Barack Obama. Executive Order 13546. Optimizing the Security of Biological Select Agents and Toxins in the United States (July 2, 2010). Available at: http://www.whitehouse.gov/the-press-office/executive-order-optimizing-security-biological-select-agents-and-toxins-united-stat (accessed May 15, 2012).
10. Animal and Plant Health Inspection Service. Agricultural Bioterrorism Protection Act of 2002. Biennial Review and Republication of the Select Agent and Toxin List. Amendments to the Select Agent and Toxin Regulations. 7 CFR 331; 9 CFR 121 (October 5, 2012).
11. Health and Human Services. Possession, Use, and Transfer of Select Agents and Toxins. Biennial Review. 42 CFR 73 (October 5, 2012).
12. U.S. officials declare researcher is anthrax killer. CNN (August 6, 2008). Available at: http://articles.cnn.com/2008-08-06/justice/anthrax.case_1_bruce-ivins-anthrax-killer-anthrax-attacks?_s=PM:CRIME (accessed June 11, 2012).
13. National Research Council (U.S.). Committee on Laboratory Security and Personnel Reliability Assurance Systems for Laboratories Conducting Research on Biological Select Agents and Toxins. National Academies Press. ebrary; Responsible research with biological Select Agents and toxins. National Academies Press. Washington, DC (2009). Accessed ar: http://site.ebrary.com/lib/pitt/Doc?id=10367629.

14. National Science Advisory Board for Biosecurity. Enhancing personnel reliability among individuals with access to Select Agents (2009). Available at: http://oba.od.nih.gov/biosecurity/meetings/200905T/NSABB%20Final%20Report%20on%20PR%205-29-09.pdf (accessed July 18, 2012).
15. Federal Experts Security Advisory Panel. Recommendations Concerning the Select Agent Program (2011).
16. Government Accountability Organization. GAO-09-574. High-containment laboratories: national strategy for oversight is needed. (September 2009). Available at: http://www.gao.gov/new.items/d09574.pdf (accessed July 18, 2012).
17. Skvorc, C., and Wilson, D.E. Developing a behavioral health screening program for BSL-4 laboratory workers at the National Institutes of Health. Biosecur Bioterror. Mar 9(1):23–29.
18. Franz, D.R., and LeDuc, J.W. Commentary: Balancing our approach to the insider threat. Biosecur Bioterror. 9(3):1–2 (2011).
19. Gronvall, G.K. Managing the Insider Threat in High-Containment Laboratories. UPMC CfBo (ed). Crossroads in Biosecurity: Steps to Strengthen U.S. Preparedness.2011) Available at: http://www.upmc-biosecurity.org/website/resources/publications/2011/2011-2009-2008-managing%2020insider-threat_crossroads.html (accessed July 2019, 2012).
20. Franz, D.R., Ehrlich, S.A., Casadevall, A., Imperiale, M.J., and Keim, P.S. The "nuclearization" of biology is a threat to health and security. Biosecur Bioterror. 7(3): 243–244 (Sept. 2009).
21. Borum, R. Psychology of Terrorism. University of South Florida. Tampa (2004).
22. Report of the Defense Science Board Task Force on Department of Defense biological safety and security program. Office of the Under Secretary of Defense for Acquisition, Technology, and Logistics. Washington, D.C. (2009).
23. United States Government Policy for Oversight of Life Sciences Dual Use Research of Concern (March 29, 2012). Available at: http://oba.od.nih.gov/oba/biosecurity/PDF/United_States_Government_Policy_for_Oversight_of_DURC_FINAL_version_032812.pdf.
24. Roos, R. H5N1 transmission experiment stirs concern. CIDRAP (November 17, 2011).
25. Malakoff, D. Breaking News: NSABB Reverses Position on Flu Papers. Science Insider (March 30 2012). Available at: http://news.sciencemag.org/scienceinsider/2012/03/breaking-news-nsabb-reverses-pos.html (accessed April 3, 2012).
26. Imai, M., Watanabe, T., Hatta, M., et al. Experimental adaptation of an influenza H5 HA confers respiratory droplet transmission to a reassortant H5 HA/H1N1 virus in ferrets. Nature. 86(7403):420–428 (June 21 2012).
27. Russell, C.A., Fonville, J.M., Brown, A.E., et al. The potential for respiratory droplet-transmissible A/H5N1 influenza virus to evolve in a mammalian host. Science. 336(6088):1541–1547 (June 22 2012).
28. Schnirring, L. NIH preparing tool to help labs cope with new DURC policy. CIDRAP News (May 1 2012). Available at: http://www.cidrap.umn.edu/cidrap/content/influenza/avianflu/news/may0112workshop-br.html (accessed May 15, 2012).
29. Wolinetz, C.D. Implementing the new U.S. dual-use policy. Science. 336(6088): 1525–1527 (June 22 2012).
30. Kwik, G., Fitzgerald, J., Inglesby, T.V., and O'Toole, T. Biosecurity: responsible stewardship of bioscience in an age of catastrophic terrorism. Biosecur Bioterror. 1(1):27–35 (2003).

31. Epstein, G.L. Preventing biological weapon development through the governance of life science research. Biosecur Bioterror.10(1):17–37 (March).
32. Public Health, Biosecurity, and H5N1 Science (2012). Available at: www.sciencemag.org/site/feature/data/hottopics/biosecurity/ (accessed on July 20, 2012).
33. Cello, J, and Paul, A.V. Wimmer E. Chemical synthesis of poliovirus cDNA: generation of infectious virus in the absence of natural template. Science. 297(5583):1016–1018 (August 9, 2002).
34. Couzin-Frankel, J. Polio Paper Sparks Criticism From Congressional Representatives. ScienceNOW (July 29, 2002). Available at: http://news.sciencemag.org/sciencenow/2002/07/29-02.html (accessed July 18, 2012).
35. Smith, H.O., Hutchison, C.A., 3rd, Pfannkoch, C., Venter, J.C. Generating a synthetic genome by whole genome assembly: phiX174 bacteriophage from synthetic oligonucleotides. Proc Natl Acad Sci U S A. 100(26):15440–15445 (Dec. 23 2003).
36. Presidential Commission for the Study of Bioethical Issues. Press Release: Presidential Commission on Bioethics Calls for Enhanced Federal Oversight in Emerging Field of Synthetic Biology (2010). Available at: http://bioethics.gov/cms/node/354 (accessed May 16, 2012).
37. Andrew, P. Presidential Bioethics Panel Gives a Green Light to Research in Synthetic Biology New York Times.
38. Screening Framework Guidance for Providers of Synthetic Double-Stranded DNA. Federal Register.
39. Gronvall, G.K. HHS guidance on synthetic DNA is the right step. Biosecur Bioterror. 8(4):373–376 (Dec. 2010).

CHAPTER 12

The Link between Biosecurity and Bioterrorism

Daniel N. Apple and Benjamin A. Perman

INTRODUCTION

Bioterrorism represents a set of terrorist actions involving the use of biological agents (etiological organisms and viruses) and biological toxins (macromolecular proteins) to deliberately inflict morbidity or mortality on a target community. Bioterrorism requires unique countermeasures to prevent attacks and mitigate their effects. In the jurisdictional limits of the U.S., the Federal Bureau of Investigation (FBI) is the lead agency for responding to and investigating acts of terrorism, including acts of bioterrorism. The FBI protects national security by developing and implementing specific countermeasures that prevent attacks, mitigate the effects of biological weapons (BW), and prevent future attacks. The Biological Countermeasures Unit (BCU) is the executive element of the FBI Weapons of Mass Destruction Directorate (WMDD), which has responsibility for developing and implementing FBI countermeasures for bioterrorism. The goal of the WMDD BCU is to develop biological countermeasures by:

1. Raising awareness
2. Threat detection
3. Decreasing vulnerability gaps

BIOTERRORISM ATTACK CYCLE

Terrorist operations are based on a "shock-action" paradigm. Like many Special Forces missions, terrorist activity relies heavily on covert logistics and operations support, close reconnaissance or infiltration, and unit-level mission planning, egress, and recovery. Terrorist action may have special requirements that depend on the specific aspects of specific forces, weapons, and targets. This is true in bioterrorism, where weapons are derived from biological agents or toxins, which introduces specific features to the model for terrorist action. Bioterrorism involves the following phased sequence or cycle of events (Figure 12.1):

1. Establishment of a covert logistics/operations support capability
2. Close surveillance of the source of weapons precursor materials
3. Infiltration of the source of weapons precursor materials
4. Theft or diversion of weapons precursor materials
5. Weapons development
6. Weapons production
7. Weapons packaging/preparation
8. Close surveillance of the target
9. Assault
10. Egress and recovery
11. Postattack target surveillance to measure weapons effects

The term "attack" refers to the set of tactical actions that support the deployment and use of the weapon, agent, or pathogen. The attack itself may involve

FIGURE 12.1 Bioterrorism attack cycle

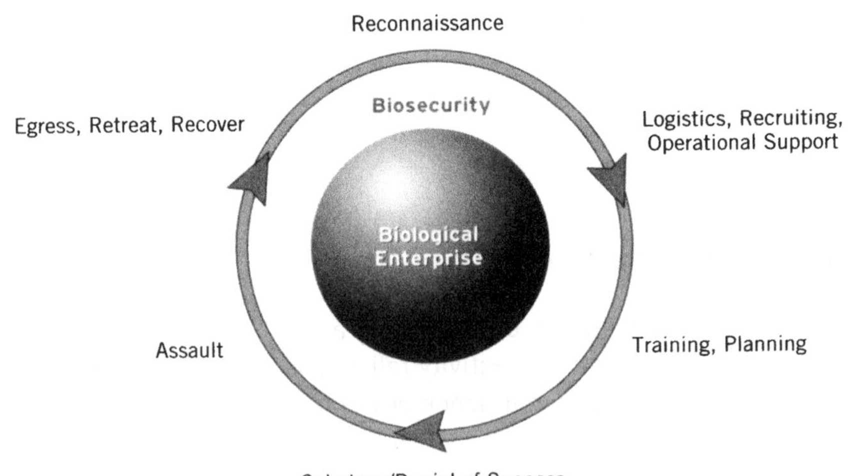

persons who have training or familiarity with terrorist operations and may not involve knowledge of the weapon or materials that have gone into the weapon. The attack model is separated into three phases: (1) acquisition, (2) development, and (3) operations, orchestrated by the organization's command and logistics branch (Figure 12.2).

Command and Logistics

Command and control are vital to the coordination of any organization. Even simple terrorist organizations or operations require coordination of separate elements.

Logistics support is at the core of all terrorist actions. Even modest operations require a source of funding and coordination of all required elements of the attack. Logistics typically include basic financial support to the operation, recruiting of operators, transportation of weapons, and other equipment.

FIGURE 12.2 Attack workflow diagram

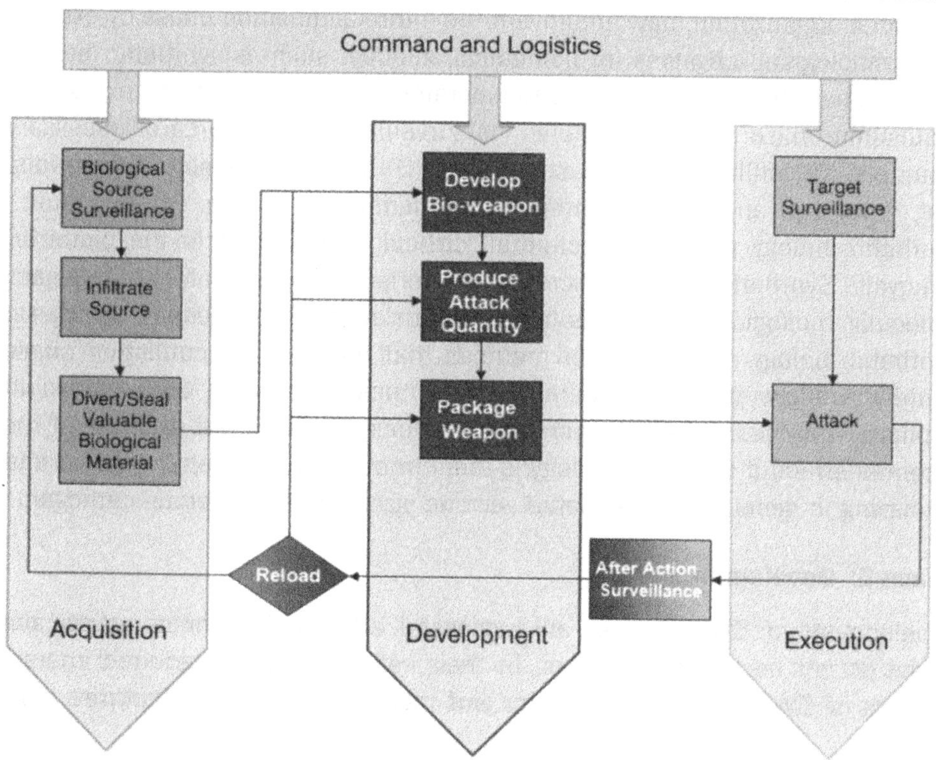

Phase 1: Acquisition

Acquisition is the series of actions required for a terrorist to obtain the necessary materials for an attack. In bioterrorism this would be removal of specific dangerous pathogens or toxins from a research or clinical lab or actively from environmental samples. Acquisition is a two-step process involving source surveillance and source infiltration:

- Source surveillance is conducted to identify a known source of a desired agent or toxin. Surveillance may also expand to cover tactical information about vulnerabilities and ingress and egress routes of relevant facilities.
- Source infiltration may result in an overt theft of valuable biological material (VBM). Infiltration leaves a prominent signature and alerts responders to the exact agent or toxin to be used in an attack. This helps responders prevent an attack and mitigate the effects of the weapon once it is released.

Removal of VBM is the tactical component of the acquisition phase. This represents a theft or diversion of VBM from a lab, clinic, or environmental reservoir.

A terrorist organization may circumvent the entire acquisition phase by exploiting technological advances in biomedical science, such as synthetic biology, that eliminate the need for access to laboratory stocks of agents or toxins. The acquisition phase typically includes (1) surveillance of a source of necessary materials, (2) infiltration of the source, and (3) theft or diversion of materials. The removal of any of these three requirements results in a "short circuit." Synthetic biology provides a potential, critical short circuit in the bioterrorism cycle. Synthetic biology describes several technologies that include legacy molecular biological techniques coupled with advances in genetic synthesis. Synthetic biology offers a set of methods that remove the acquisition phase entirely by allowing would-be terrorists to synthesize potential VBM that would ordinarily only be located in laboratory locations. The simplest example of the potential effect of synthetic biology is the introduction of genetic material that is missing in genetically attenuated vaccine strains of a dangerous pathogen.

Phase 2: Development

Development of BW precedes all successful attacks. Infectious agents and toxins do not necessarily possess, in their native form, the required characteristics of BW. To assure morbidity and mortality in victims, features must be deliberately engineered into the agent or toxin. Further, many potential BW pose specific tactical limitations that must be controlled or overcome by would-be terrorists. Some examples are: required vectors, routes of infection, lethal doses, resiliency, and a variety of etiological and epidemiological factors. Development tends to require significant expertise in microbiology and related fields as well as specific laboratory facilities. For example, one may be able to synthesize smallpox from another related pox virus by using freely available genetic data.

Phase 3: Operations

Production of the agent or toxin for an attack is perhaps one of the most difficult and time-consuming processes. It involves a significant scale-up of proof-of-concept trials in the development stage. The process of scale-up may involve significant reengineering of the agents or toxins to facilitate or enhance the production of larger quantities.

Packaging is the loading of a weapon in a delivery device, munition, or vector. Without the proper loaded delivery device, dissemination of even large quantities of agents or toxins can be rendered ineffective because they are not delivered in a manner that causes infection, are subjected to environmental conditions that weaken them or fail to deliver a lethal dose.

After action, the terrorist will conduct surveillance to observe, note, and, later, improve weapons effects. Terrorists with the capability will also try to learn from the results of the attack by studying the epidemiology of the attack and the etiology of the weapon to improve on various development processes that may have failed or can be improved. The afteraction surveillance is very similar to criminal and epidemiological investigations that will be undertaken by responders to assist victims.

After an initial attack, terrorist units can assess weapons' effects and tactics to determine if a follow-on attack is possible or if plans require modification. If they are changing weapon type, the terrorist unit then returns to either the initial surveillance phase on a different target or to any other point in the cycle, based on the reload requirements for another assault phase. This model describes a "reload" pathway. Some reload pathways will lead back to refinements of the weapon development effort or introduction of more of the same weapon.

THE BIOTERRORISM-BIOSECURITY NEXUS

Countermeasures deployed against BW and their use in acts of terrorism are derived largely from basic federal law enforcement and investigative authorities, coupled with specific criminal prohibitions on the development of any BW, possession of certain biological agents and toxins, and acts of terrorism involving weapons of mass destruction (WMD) themselves. The Biological Weapons and Toxins (BWAT) Statute (18 USC 175) is the implementation of the Biological Weapons and Toxins Convention in U.S. law.[1] It makes the development, possession, and use of any biological weapon a felony and has recently been modified to include unlawful position of biological select agents and toxins (BSAT), defined in the Select Agent Rules 42 CFR 73, 9 CFR 121, and 7 CFR 331 (see Chapters 2 and 7), without the requirement to establish intent to either develop or use them as weapons.[2,3] Other facets of the criminal code are important to investigating and ultimately prosecuting biological terrorism cases, but the BWAT statute is intended to counter biological terrorism specifically at the acquisition and development phases, where FBI WMD equities have the specialized resources and capabilities to act. The BWAT statute, coupled with the select agent rules, is the cornerstone of the countermeasures driving other FBI WMD outreach efforts and operations.

Countermeasure Requirements

Countermeasure requirements are derived from vulnerabilities revealed in an analysis of the bioterrorism attack model (Figure 12.2) and authority in the prevailing statutory doctrine. A countermeasure blocks or restricts terrorist action by interrupting one or more of the individual processes in the attack model described above. During the development of countermeasures requirements, three aspects of the countermeasure must be considered to assess the relative cost/benefit of the countermeasure: (1) where it exists in the attack model, (2) whether its placement in the cycle presents special difficulties, and (3) which tactics, techniques and procedures will be necessary to implement effective countermeasures.

Countermeasures will have a greater or lesser impact on blocking or restricting terrorist action by their position in the attack model because they lie on the critical path of the terrorist workflow. In other words, the terrorist has no way to progress through the process if he is blocked by a countermeasure. Investment in countermeasures on the critical path will tend to have a greater benefit than those that are parallel to the work-flow.

Countermeasures in some parts or phases of the attack model will tend to be easily implemented by the nature of the specific characteristics of individual processes in that phase, but others will not. For example, control of available technologies that could be used by terrorists is difficult in the free and transparent scientific enterprise that exists in the U.S. and much of the developed world.

Countermeasures require different tactics, techniques, and procedures (TTPs) that match the TTPs and procedures used by the terrorist at each phase in the attack model. The acquisition portion involves TTPs derived from intelligence/counterintelligence operations and the medical/biomedical research communities. The development phase is dominated by TTPs from the medical/biomedical research fields, with significant contributions from more traditional special warfare skills. The command/logistics phase draws on command and control, logistics, and administrative skills. Finally, the operational phase is dominated by traditional special warfare and counterterrorism TTPs. Table 12.1 shows that TTPs can be used to develop four groups of countermeasures, including threat identification, biosecurity, technology security, and target security, that relate to the four distinct phases of the attack model.

Countermeasure Development

Biological terrorism countermeasures develop through a spectrum of increasing maturity that includes three major steps: relationships, partnerships, and systems. All countermeasures are targeted towards the eventual development of fully functional systems. However, the reality of countermeasure development is that some countermeasures may attain a high degree of effectiveness in the earlier steps of their development and some, due to cost or schedule constraints, may never achieve full maturity.

TABLE 12.1 Countermeasure types related to attack cycle phases

Attack Cycle Phase	Countermeasure Type
Command/logistics	Threat identification
Acquisition	Biosecurity
Development	Technology security
Operations	Target security

THE BIOLOGICAL CRIMINAL CONTINUUM

The biological criminal continuum (BCC) is the BCU's fundamental understanding of criminal/terrorist biological attacks. Like the attack model described above, the BCC is composed of three phases under the command and logistics components of the terrorist operation (Figure 12.3). The criminal/terrorist organization provides the organizational control and motivation to commit the criminal act and involves the three other components: acquisition, development, and execution. The organization or criminal decides based on operational requirements how to carry out each component. The individual actions comprising each component are either performed sequentially or simultaneously. The BCC can be composed of one "lone wolf" threat that undertakes all necessary components. The component view of the BCC helps to illustrate why a "lone wolf" threat is a fairly unique individual who possesses an unusually diverse set of skills and motivations.

The BCU approaches the BCC using three components because each component is facilitated by a unique capability. The human capital for each capability

FIGURE 12.3 Biological criminal continuum

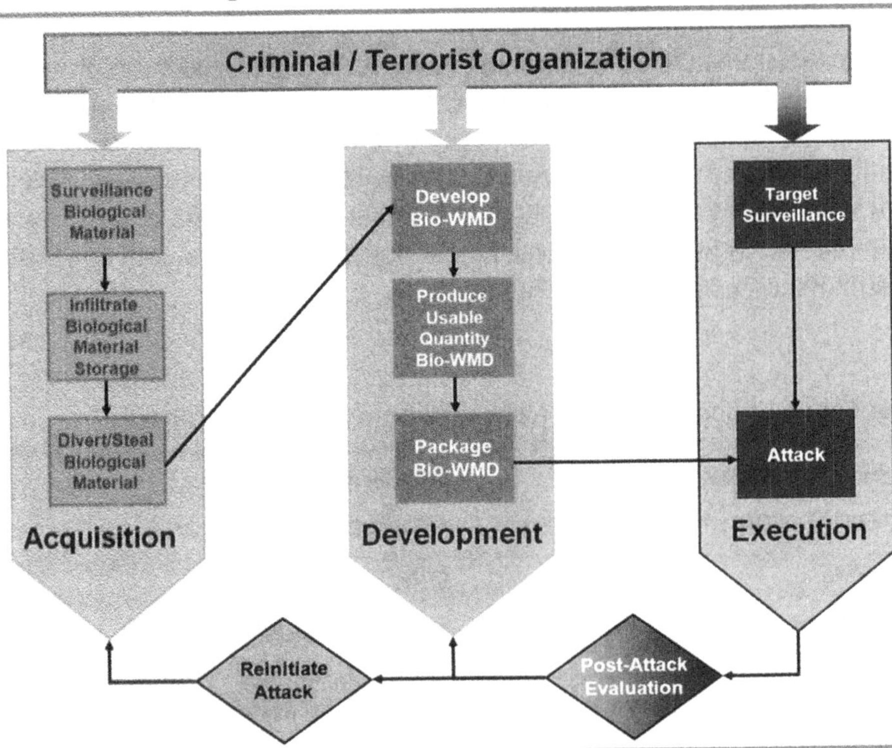

is composed of specialists who have unique capabilities or knowledge in the required individual actions. Specialists are persons with varying degrees of access to valuable biological materials that could be used as biological weapons.

There may be overlap among certain specialists and components. For example, in the acquisition component, a biomedical scientist who has access to the biological material may be an active participant in advancing the development of the material into a pathogenic weapon for dissemination. Alternatively, there could be a biomedical scientist who acquires the biological material and a different scientist who develops the pathogen for dissemination. There are several possible threats that may operate in the acquisition component of the BCC. The threats are organized into insiders and outsiders (Chapter 6):

1. Acquisition is a series of actions required for a criminal/terrorist to obtain the necessary materials for an attack that include:
    - Surveillance of biological material: what is needed for an attack and where it can be acquired—surveillance in this case is directed against the institutions that own, house, or transfer materials; moreover, it is directed at uncovering specific information about materials and vulnerabilities
    - Infiltration of biological material storage; actions required to access either directly or indirectly (manipulating persons with access)
    - Diversion/theft of biological material; final action in the process of acquiring materials for use as weapons
2. Development is a series of actions required for a criminal/terrorist to advance the development of the biological WMD. Many potential bioweapons pose specific tactical limitations that must be controlled or overcome, such as required vectors, routes of infection, lethal doses, resiliency, and a variety of etiological and epidemiological factors. Infectious agents and toxins do not necessarily possess, in their native form, the required characteristics of bioweapons. To produce morbidity and mortality in victims, these features must be deliberately engineered into the agent or toxin. These actions require a specific set of technical specialists. Synthetic biology and other technologies may be leveraged to enhance development and may require a special countermeasures focus. Amateur biologists may represent another important caveat, because specialists with the requisite skills may acquire skills outside of traditional academic or industrial research settings. Successful biological weapons attacks may require unusually large quantities of material. A criminal enterprise that needs large quantities of material will have

very specific requirements for equipment and expertise and may need to develop new protocols for scaling up production. Successful attack also requires the correct use of dispersal techniques or vectors. Packaging requires a unique set of skills distinct from the previous two actions in the development component of the BCC.

3. Execution is a set of actions required for a criminal/terrorist to execute a biological WMD attack. Target surveillance is necessary in order to have maximum effect. Factors such as weather, deployed countermeasures, and animal reservoirs used as disease vectors are elements that will shape the characteristics of the attack. A postattack evaluation may be undertaken by the criminal/terrorist organization to study the weapon's effects on the target area and population in an attempt to gather information that could be used to validate weapon development and attack tactics or identify refinements that are needed to improve the weapon or tactics. Postattack evaluation contains aspects of the development and execution components. Reinitiation of attack occurs after the initial attack. Criminals/terrorists can assess bioweapon effects and tactics to determine if a follow-on attack is possible or if plans should change to enhance effects or circumvent countermeasures. Reinitiating the attack may require that acquisition be revisited to obtain new or different materials. Reinitiation may require refinement of the weapon in the development component. As such reinitiation contains aspects of the acquisition and development components.

CONCLUSION

Theft and diversion of valuable biological materials in the acquisition mode of the biosecurity continuum require that threats follow one of two possible generalized pathways in and out of a biosecurity system. A breach of the system results when the threat completes one entire circuit of the system, resulting in valuable biological materials being removed from the accountability system inside the laboratory. The "inner sanctum" in biosecurity is the repository of information and materials that is protected from loss and is shown as a red-dashed perimeter in Figure 12.4. There may be multiple security enclaves and perimeters inside of other perimeters. In Figure 12.4 this concept is shown by the dashed line institutional perimeter, which is analogous to the outer perimeter of a building or fence line around a campus. Countermeasures designed to defeat insider and outsider threats are positioned along the breach pathways: arrows for insiders heading out and outsiders heading in. Some countermeasures are common to insiders and outsiders, such as access control systems

FIGURE 12.4 Biosecurity operational concept

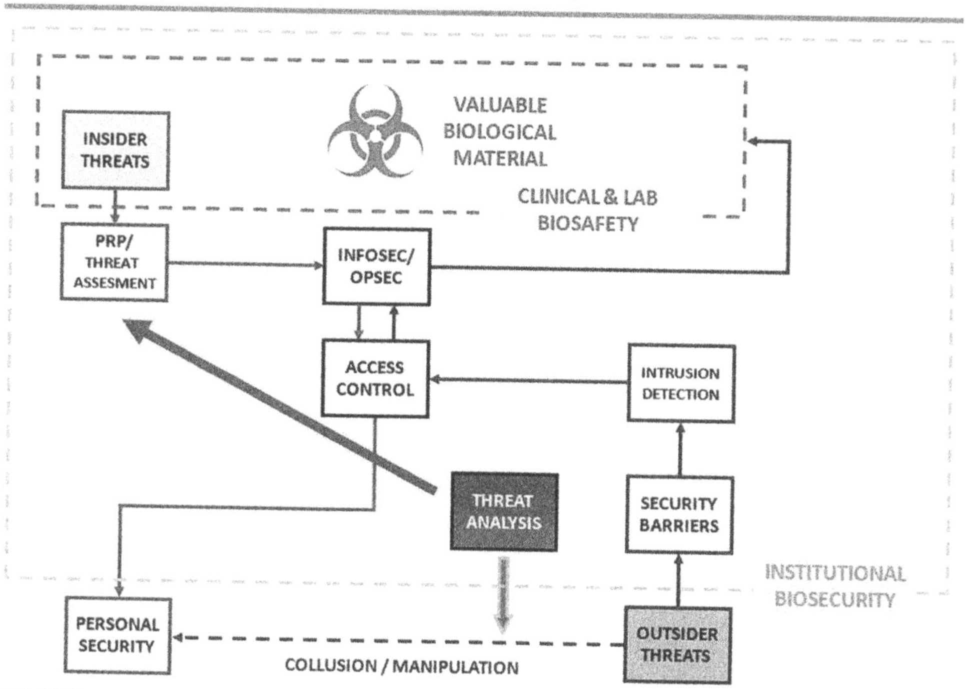

and information and operational security (discussed in detail in Chapter 5). Others are specifically tailored to the threat. Barriers and intrusion detection systems mitigate outsider threats, and reliability and personal security systems are specifically designed to mitigate insider threats. The most effective insider threat countermeasures involve threat analysis. Threat analysis also is the most effective way of dealing with collusion between outsiders and insiders and manipulation of insiders by outsiders. Figure 12.4 shows that collusion and manipulation by an outsider threat are critical for theft and diversion of valuable biological material because they enable infiltration of the institution and its biosecurity systems, a prerequisite for theft and diversion of valuable biological material in the acquisition phase of the biosecurity continuum. These countermeasures are covered in Chapter 6.

References

1. Arms Control Association. Biological Weapons Convention. Convention on the prohibition of the development, production, and stockpiling of bacteriological (biological) and toxin weapons and on their destruction (1972). Available at: http://www.armscontrol.org/treaties/bwc

2. U.S. Department of Agriculture. Animal and Plant Health Inspection Service. 2 APHIS Select Agents Regulations. Federal Register. 7 CFR Part 331 and 9 CFR Part 121 (2008).
3. Department of Health and Human Services. Centers for Disease Control. CDC Select Agents Regulations. Federal Register 42 CFR Part 73 (2008).

CHAPTER 13

The Importance of Engagement and Education for Effective Biosecurity

James Revill and Catherine Jefferson

INTRODUCTION

Establishing an effective biosecurity program is unlikely to be achieved by physical measures such as guards, guns, and gates alone. While these physical components are important, effective biosecurity also requires engagement and education on issue relating to dual-use research. This chapter explores global biosecurity-related engagement and education initiatives and addresses challenges in promoting biosecurity education, importance of biosecurity education, how education and engagement initiatives can be enhanced for future initiatives, and mechanisms for institutional change.

OVERVIEW OF ENGAGEMENT AND EDUCATION

There has been a greater need for biosecurity education and engagement programs over the last decade. In particular, several international, regional, and like-minded collectives of states have placed emphasis on programs for scientists dealing with a range of dual-use (see also Chapters 7 and 11) technologies.[1] Those dealing with the life sciences are no exception; indeed, in part because of the changing capacities and geography of life science research, this sector is more important, and there have been a number of calls for greater engagement and education. Most recently, this has been evidenced by the states parties to the Biological and Toxin Weapons Convention, which agreed, in December 2011, by consensus on: "... the value of national implementation measures, as appropriate, in accordance with the constitutional process of each State Party, to ... promote the development of training and education programmes for those granted access to biological agents and toxins relevant to the Convention and for those with the knowledge or capacity to modify such agents and toxins;"[2]

Similar support for education has been expounded by the G8 Global Partnership,[3] the European Union,[4] and the Organization for Economic and Cooperation and Development (OECD),[5] as well as a number of national statements and proposals from NGOs and academics.[6]

Yet, despite the interest in engagement and education from a number of different stakeholders, interest among scientists remains relatively modest. Indeed, what has emerged is a global patchwork of surveys and national or transnational education and engagement projects supported by NGOs and academics.[7] Examples of educational programs are illustrated in Figure 13.1 with particularly salient programs highlighted below:

- In Australia, the Department of Health and Ageing has developed "an education and awareness-raising program ... to promote recognition and understanding of the security-sensitive biological agents (SSBAs)."[8]
- The Public Health Agency of Canada (PHAC) has been working on the "development and delivery of a curriculum for a university-level accredited pilot course on applied dual-use biosecurity" in collaboration with Bradford University.[9]
- In Japan, the National Medical Defence College has "developed an online learning module in applied dual-use biosecurity education" in collaboration with the University of Bradford. [10]
- According a 2012 paper, "over the past 10 years the U.S. government, academic and scientific institutions, and industry and nongovernmental organizations (NGOs) have worked together to develop, implement and raise awareness about biosecurity and biosecurity education."[11] Such activities include a survey of U.S. university course content led by AAAS,[12] the establishment of an NSABB Working Group on Outreach and Education,[13] the Federal Bureau of Investigation (FBI)-led Amateur (Do-It-Yourself) Biology Initiative, and the FBI Synthetic Biology Program.
- The University of Nairobi "plans to conduct an online biosecurity education course for the life science community in East Africa." [14]
- The Korea Human Resource Development Institute for Health and Welfare (KOHI) is "develop[ing] an education program for public officers and researchers working in the field of biomedical science." [15]

FIGURE 13.1 Locations of major education and awareness raising projects around the globe

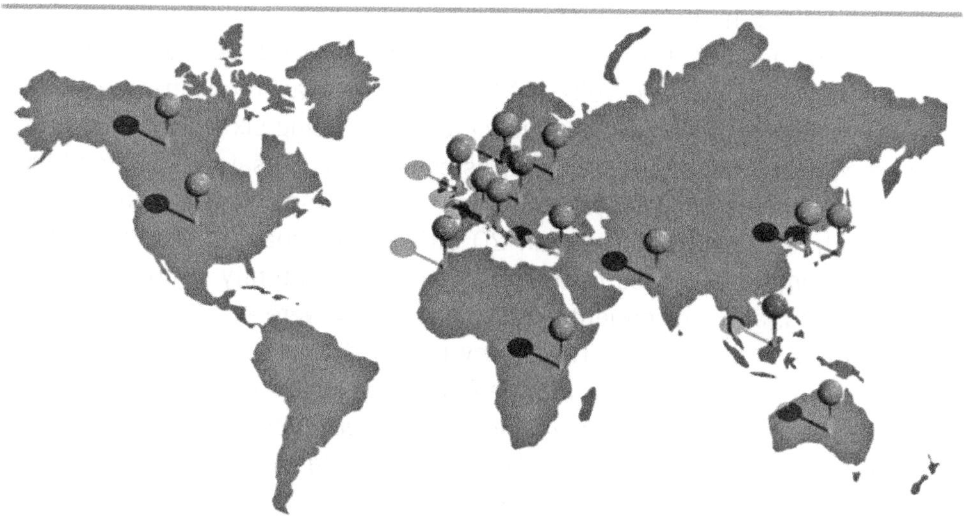

- The Swiss government used a "biology for peace" brochure as well as a series of awareness-raising seminars conducted by experts to begin the process of sensitizing researchers.[16]
- National Academy of Sciences of Ukraine (NASU), along with academics from the Universities of Exeter and Bradford, arranged "an interactive seminar in 2007 on biosafety education including the relevance of the BWC for Ukrainian scientists and students."[17]
- A Pakistani-based initiative led by the Quaid-e-Azam University (QAU) involved a survey and the subsequent development of a course on bioethics, biosafety, biosecurity, and dual-use education.[18]
- Swedish experts have established a network as part of a process of building an educational program covering bioethics, biosafety/biosecurity, and dual-use. [19]
- The Italian National Board of Biologists (CBUI) agreed to include dual-use-related issues in the biology curriculum at bachelor, master, and doctoral levels.[20]
- A survey of Israeli university life science courses for biosecurity or dual-use content has been conducted.
- A nationwide initiative is underway in Morocco to reach out to life science students and raise awareness of dual-use issues.

In addition to these national or locally orientated initiatives, there have been regional attempts to gauge the extent to which students are exposed to biosecurity-related issues through university-level training, primarily in Southeast Asia and Japan. In one project, Bradford Disarmament Research Centre (BDRC) and the Landau Network Centro Volta (LNCV) conducted a European-wide survey of online syllabi of university life science courses, using six search terms: (1) codes of conduct, (2) dual-use, (3) arms control, (4) biosafety, (5) bioethics, and (6) biosecurity. The analysis included 142 courses from 57 universities in 29 countries speaking more than 20 different languages. After the initial survey was complete, course instructors had the opportunity to confirm or revise the results, and subsequently BDRC and LNCV were able to discern their attitudes towards biosecurity. The survey provided an illustrative snapshot of course content and attitudes within the European region. The results indicated that although bioethics and equivalent courses were prevalent, there was less evidence of specific biosafety modules and very little evidence of biosecurity, dual-use, or arms control (Figure 13.2).

The qualitative results developed from discussions with course instructors and coordinators provided an often-neglected perspective on the attitudes stemming from colleges and universities. For some, it was apparent, the integration of dual-use issues was not problematic, with one instructor stating, "I will take profit of your enquiry and integrate these issues in the introductory course starting with this year;" for others education was "not [their] cup of tea." However, the most common response was that inserting security-related material into an already "busy" curriculum would be difficult and there was a lack of resources to do so.

FIGURE 13.2 Results from the European Survey of life science curricula

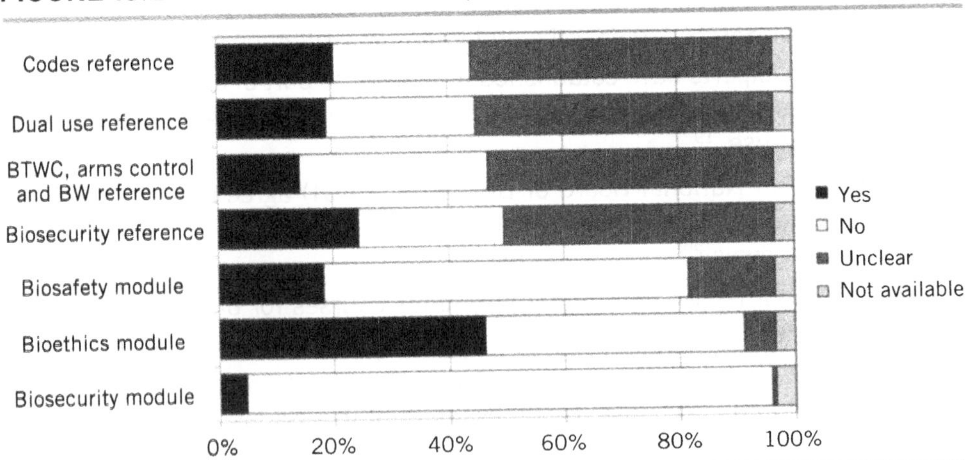

OTHER INTERVENTION POINTS AND ACTORS

Aside from the traditional educational activities conducted at universities, a number of other stakeholders and mechanisms for engagement or "intervention points" for effective biosecurity engagement and education have been identified. In terms of actors, there is broad acceptance that a range of different stakeholders at different levels of authority need to be engaged to achieve biosecurity. As Vahid Majidi, Assistant Director, Weapons of Mass Destruction Directorate, Federal Bureau of Investigation, indicated in a statement to the Senate Committee on Homeland Security and Governmental Affairs in 2011, "The FBI recognizes additional stakeholders from federal, state, and local government and nongovernment agencies have roles, responsibilities, and resources to ensure safe and secure research with biological pathogens and toxins. Inclusion of these stakeholders and coordinated efforts are also vital for biosecurity."[21]

In terms of other intervention points, examples include biosecurity as a consideration in funding criteria, the development of codes of conduct, addition of biosecurity-related textbook content, enhanced law enforcement outreach, engagement through academies of science, and working with journal editors. These approaches cumulatively can be mutually reinforcing, encouraging a "cradle to grave" awareness of dual-use issues and by implication building stronger foundations for future biosecurity-related activities. There is evidence of activity around some of these intervention points.

Funders

In the U.K., a major funding agency for bioscience research released a joint policy statement in 2005 on "managing risks of misuse associated with grant funding activities," in which they indicated that they have introduced, inter alia, a question on application forms asking applicants to consider the risks of misuse associated with their proposals.[22] Similar steps have been taken by the European Union (EU), which now has dual-use as a checklist heading for the Framework Program 7 ethics review.[23] In the U.S., the National Academies of Sciences (NAS) report on dual-use education referred to the "growing interest in expanding education about dual-use issues, such as the proposal to require such education for all federally funded life scientists in the United States." However, it was not until 2012 when the bungling of the dual-use debate surrounding two bird flu related experiments resulted in new rules that require federal agencies to systematically screen funding proposals for "dual-use research of concern."[24]

Codes of Conduct

In 2005, state parties to the Biological Toxin and Weapons Convention (BTWC) focused on the topic of promoting "common understanding and effective action ... [on the] ... content, promulgation, and adoption of codes of conduct for scientists." This meeting recognized that codes of conduct: "... can support the object and purpose of the Convention by making a significant and effective contribution ... to combating the present and future threats posed by biological and toxin weapons, as well as by raising awareness of the Convention, and by helping relevant actors to fulfil their legal, regulatory, and professional obligations and ethical principles."[25]

Following this meeting there was a much greater understanding of the different types of codes and the different functions they can play as well as a number of efforts to develop codes of conduct. Examples include the work on the Inter Academy Panel, which released a statement on biosecurity consisting of a number of key principles signed by academies around the globe.[26]

Textbooks

The inclusion of material on biosecurity or dual-use in academic textbooks for biology students remains limited, despite state parties to the BTWC calling for "[its] inclusion in textbooks and in medical, scientific, and military educational programmes of information dealing with the prohibition of bacteriological (biological) and toxin weapons and the provisions of the Geneva Protocol" since the Second Review Conference.[27] This is particularly disappointing, as inserting biosecurity-related content into core textbooks used by undergraduate and postgraduate life science students represents one approach that ensures a degree of awareness of the issues involved. Moreover, textbook authors seeking to update past editions of introductory books for biology courses are receptive to including material. For example, in a correspondence with Robert W. Bauman, the author of several editions of the text Microbiology with Diseases by Taxonomy, he stated, "I plan on expanding the content of the section on biological weapons to indicate the illegality of these weapons. I thank you for bringing this matter to my attention. I think it is wise to enhance greater understanding of this important issue." Clearly, inserting materials into textbooks as a standalone measure is unlikely to be successful; however, taken alongside a series of other measures, this is an area in which further advances could be forthcoming.

Law Enforcement Outreach

A novel approach to biosecurity-related engagement can be found in the work of law enforcement agencies, particularly in the U.S., where there have been various FBI-led outreach and engagement initiatives. These include the FBI Academic Biosecurity Workshops, which are an ongoing series of meetings intended to inter alia "improve the cooperation among law enforcement agencies and research institutions to mitigate potential biosecurity issues that may affect public health and safety"[28] and FBI efforts to develop a partnership with the so-called "DIY bio" community "in order to garner their assistance in preventing, detecting, and responding to incidents of misuse, particularly for nefarious purposes."[29] This more delicate approach to engagement appears to have made some progress in building bridges between two communities and reportedly assisted in "com[ing] up with strategies that make sense to both communities to mitigate ... risks and gaps."[30]

Academies of Science

Academies of science, societies, and associations have a clear role to play in supporting the engagement with as well as the development, promulgation, and adoption of biosecurity-related education. Indeed, in the absence of an international structure similar to the International Union of Pure and Applied Chemistry (IUPAC), which has played a valuable role in the evolution of the Chemical Weapons Convention, academies of science, their equivalents, and their international manifestation in the Inter Academy Panel are increasingly important.

CHALLENGES MOVING FORWARD

Despite a concerted effort by different stakeholders, progress in education and engagement has been hampered by a number of barriers. Certainly, at the level of life science education, several studies have reached conclusions that security-related material has a limited presence in the university curriculum for the following reasons:

- An absence of space in the existing curricula
- An absence of time and resources to develop new curriculum materials
- An absence of expertise and available literature on biosecurity education
- The perceived irrelevance or limited relevance of security-related issues[31]

Perhaps of greater importance are the underlying sensitivities triggered by efforts of the security community to "educate" and "secure" the life science community and the suspicions that such initiatives can generate. Certainly, those involved in biosecurity-related education in Italy intimated as much, suggesting: "discussions on biosecurity and dual-use can be sensitive for life scientists ... It has been important to adopt an approach that emphasizes the active role of scientists in creating a culture of responsibility ... and avoids creating prejudices and generating a climate of suspicion towards science and research." [32]

This has been more baldly stated by one member of Sandia National Laboratories, who posited:

> *"One of the most significant obstacles is overcoming the impression —generally held by bioscience researchers who are not accustomed to security procedures—that biosecurity is intrusive, counterproductive, restrictive, or insulting. Creating a "biosecurity culture" by illustrating how biosecurity can be useful and not overly burdensome is one of the primary tasks to be accomplished by biosecurity system designers."*[33]

The sensitivities of the life science community have had some effect in slowing government initiatives and programs established by states, which in some cases have dwindled due to a lack of response and problems in gaining traction among the wider scientific community.[34] Certainly, the U.K. government noted that "... there are still considerable difficulties in convincing some members of the academic community that oversight and awareness ... are issues deserving attention and action."[35] The lack of success is in part due to difficulties in justifying sustained attention to abstract topics such as education and engagement, which remain at best peripheral to traditional arms control issues.

WHY BOTHER WITH BIOSECURITY-RELATED EDUCATION AND ENGAGEMENT?

Despite the difficulties encountered in pursuing engagement and outreach activities, efforts to inculcate life scientists with a degree of awareness remain important for a number of reasons. In the introduction to the book Education and Ethics in the Life Sciences, Brian Rappert posited a number of differing perceptions of the role of biosecurity-related education for life scientists. For some, education was argued to be perceived as a necessary prerequisite "for other security-related activities to be undertaken." For others, education was about "deficiency correcting" and thus was seen as a means "to correct that ignorance." For yet others education was viewed as a means of keeping arms control agreements meaningful; for yet other education was understood as a catalyst or stepping-stone for moving ahead on other matters related to biotechnology.[36]

Rappert's analysis suggests that education and engagement are likely to be disparately conceptualized by different actors and to mean different things to different people. While one can identify valid reasons for further education among scientists, they are "necessarily confined to a fairly high level of abstraction."[37] To move beyond abstraction, it is argued that two roles for education should be the focus: first, biosecurity-related education leaves students more apt to recognise biosecurity issues and react accordingly. Indeed, drawing from the work of Hoffman and Thomson on environmental education,[38] it is possible to envisage the dual-use education process as including three critical components: awareness, leading to understanding, which in turn creates the potential and capacity to recognise and respond appropriately.[39]

Secondly, dual-use education and engagement provide the foundations for a more informed discussion and debate on issues intersecting science and security. The polarized and "bungled" debate surrounding the H5N1 research illustrates the difficulties that can arise between different communities[40] and point to a lacuna in interactions between life science researchers and security officials.[41] Engagement and education, in this sense, should be a two-way process that serves not only to inform scientists of the concerns of the security community and their legal responsibilities but also informs those in the security community of the realities and limitations of biotechnology and,, perhaps more importantly, the suitability of regulatory proposals. This could be particularly valuable in moving beyond the confusion and hype that is sometimes attributed to concepts such as "synthetic biology."

Further, the H5N1 controversy points to a compelling need to both educate scientists in security *and* educate security experts in science, as part of a genuine two-way process of engagement and education. This process of engagement and reciprocal education between the scientific community and nonscientists is not a requirement that is unique to the dual-use issue. Interaction between forensics and legal communities has also been the subject of debate, not least because, as one expert in a U.K. parliamentary debate suggests, "there is a real danger that if lawyers do not understand expert evidence, then it will not be properly tested."[42] Concerns over the legal community's understanding of the science behind forensics have been raised as an issue in both the U.K. and the U.S. A U.S. National Academies Report suggested that there was a need to "educate the users of forensic science analyses, especially those in the legal community. Judges, lawyers, and law students can benefit from a greater understanding of the scientific bases underlying the forensic science disciplines and how the underlying scientific validity of techniques affects the interpretation of findings."[43]

AUK-based project described as "lowering the drawbridge" suggested that it would be important for "both legal professionals and forensic scientists to understand the vernacular and nomenclature of both science and law, as well as the working practices and customs of each group of practitioners."[44] To date it is difficult to gauge the extent to which a greater awareness of forensics among the legal community has enhanced legal process; however, anecdotal evidence suggests that: "judges and lawyers have gotten smarter on the science ... [and]... have learnt a lot from this process and are better placed to translate the science into usability within the framework ... [this] makes the consumer smarter and requires the scientists step up their game in the presentation and validation of work... Biases get called out and research held to high standard and methods have to be justifiable."[45]

It is, of course, inappropriate to extrapolate too much from the legal experience. But at the same time it is possible to envisage a number of benefits from making education a two-way process in which communities become better able to discuss issues.

MOVING BIOSECURITY ENGAGEMENT AND EDUCATION FORWARD

Biosecurity education and engagement are not a silver bullet for biosecurity. However, they are important components in the process of achieving effective biosecurity in the future. There are a number of issues to consider. First, it may be useful to develop a more holistic approach to biosecurity and build responsible conduct of research. Second, it may be useful to think about how biosecurity efforts can be supported by institutionalized changes that mutually reinforce security-related messages. Last, yet perhaps most essential, it will be important to look at creating incentives for scientists to deal with this issue in greater depth.

Holistic Approach

Several education programs have made significant progress in assimilating a number of different topics, such as "biosecurity" and "dual-use," as part of a holistic approach. In the case of one initiative in the University of Vienna, "biosafety, biosecurity, and the safe handling of materials are taught through an integrated approach, drawing on issues of ethics and GMOs." An "interdisciplinary biosecurity" program has been developed in Italy, which "incorporates risk assessment, biosafety practices, and environmental implications." In Pakistan, the module "bioethics, biosafety, biosecurity, and dual-use education" is explicit in its holistic approach to teaching, and combines technical training in laboratory biosafety with the ethics and moral philosophies underpinning biosecurity.

A holistic approach that integrates biosecurity into a broader range of teaching issues is likely to generate greater traction among the life science community and represents a mechanism to assimilate biosecurity into the broader teaching curriculum. This approach also discusses benefits of biotechnology in dealing with the threat of diseases both natural and deliberate. Certainly, the Vienna experience has cautioned against "overstat[ing] the risks posed by biology without acknowledging its inherent benefits, as seen in developments in new technologies to counter bioterrorism and biowarfare."[46]

Sustainability

Finally, any initiative that seeks to deal with the human dimension of biosecurity will need to consider how it can be made sustainable, not just financially but also in terms of ensuring "enduring value." Ensuring that initiatives have enduring value is particularly important when engaging in security and safety training, dual-use education, or the development of codes of conduct, as they all require a message to "stick" in order to be effective. Moreover, such an approach avoids the cyclical process of project-based work that episodically reiterates the same message to different audiences or different generations. Accordingly, efforts need to consider how to build a long-term relationship between the life science and security communities rather than seeing this process of engagement as a one-night stand. There are a number of possible mechanisms that could be considered to achieve enduring value, such as the development of a loose alumni network through which materials and opportunities are shared and building in a small postproject evaluation fund at preproject stage.

Incentivizing Scientists

In addition to a holistic approach to education and an institutional shift, in order for the community of practice to make the effort to engage and invest time and resources on taking biosecurity seriously, one key ingredient will be some form of incentive. Indeed, some form of incentive will be essential if the perception of biosecurity is to evolve from being seen as "intrusive, counterproductive, restrictive, or insulting" to a valuable component of the broader process of professional training and a characteristic that is desirable to employers. Part of this requires framing engagement and overtures to the scientific community in a positive manner, designating stakeholders as part of the solution rather than part of the problem, something that may also serve to reduce initial resistance. It will also require the security community to be cognizant of—and sympathetic to—the differing priorities and interests of scientists as well as the constraints endured by stakeholders in terms of time and resources.

Yet beyond such general principles, there is no one silver bullet to incentivize greater biosecurity-related education and engagement; rather it is likely that differing incentives will need to be dangled at different levels. In the absence of support for mandatory biosecurity education at the level of university education, one approach could be to offer a biosecurity component as an explicit part of an accredited module or part of a broader course on responsible research or research ethics. Such an approach may serve to encourage greater numbers to undertake such a course, particularly if it proved to enable stimulating discussion on responsible research (rather than being a lecture on the BTWC) and if there was a demand for graduates who were more aware of biosecurity and safety issues.

At the level of the industry, incentives could come in many forms. For some organizations it may be fruitful to appeal to a sense of corporate social responsibility, and already some organizations such as Astra Zenica are taking the lead in promoting their biosecurity-related activities as part of their research ethics. The achievement of a critical mass of corporate leaders from the biotech industry could be influential in setting a trend in which biosecurity is taken seriously and made visible. It is notable that as well as improving an organization's biosafety and biosecurity procedures, compliance with CWA 15793 also provides a number of other advantages such as raising reputation, improving prospects when bidding for contracts or funding, and possibly lowering insurance premiums. In addition, adoption of such standards could benefit international cooperation and increase commercial opportunities for life science institutions in countries where biosafety and biosecurity legislation is underdeveloped. By offering tangible benefits to an organization, international management standards could provide an attractive driving force for institutional change. However, greater international recognition and adoption are needed. One possible way forward could be the development of an ISO standard on biorisk management that builds on CWA 15793. By appealing to business interests, international management standards provide a means to incentivize institutions to ensure that biosafety and biosecurity form part of an effective infrastructure.

CONCLUSION

As this chapter has indicated, achieving effective biosecurity cannot be achieved through guns, guards, and gates alone; rather it requires processes of engagement and education between the life science community and the security community. This is important, first because such measures could make life scientists better able to recognize biosecurity issues and react accordingly; and second, because measures such as dual-use education and engagement provide the foundations for more informed discussion and debate between two different communities on issues that lie at the intersection of science and security, which is likely to be increasingly important as biotechnology continues to advance.

References

1. See, for example, Pearson, G.S., and Mahaffy. P. International Union of Pure and Applied Chemistry Committee on Chemistry Education, Outreach, and Codes of Conduct to Further the Norms and Obligations of the Chemical Weapons Convention. Pure Appl. Chem. 78(11) pp. 2169–2192 (2006).
2. BWC. Final document of the Seventh Review Conference of the BWC. BWC/CONF.VII/7 (January 13, 201. Accessed at: http://daccess-ods.un.org/access.nsf/Get?Open&DS=BWC/CONF.VII/7&Lang=E.
3. G8 Summit. The Global Partnership: Assessment and Options for Future Programming, Deauville (May 27, 2011). Accessed at: http://www.g20-g8.com/g8-g20/g8/english/the-2011-summit/declarations-and-reports/appendices/g8-global-partnership-assessment-and-options-for.1354.html.
4. European Commission. *Green Paper on Bio-Preparedness.* COM 399 final. Commission of the European Communities. Brussels (July 11, 2007). Accessed at: http://ec.europa.eu/food/resources/gp_bio_preparedness_en.pdf.
5. OECD. Best Practice Guidelines on Biosecurity for BRCs (2007). Accessed at: http://www.oecd.org/document/36/0,3746,en_2649_34537_38777060_1_1_1_1,00.html.
6. See, for example, National Research Council. Challenges and Opportunities for Education About Dual Use Issues in the Life Sciences (2010).
7. See, for example, Australia, Canada, Japan, New Zealand, Republic of Korea, and Switzerland (on behalf of the JACKSNNZ) and Kenya, Pakistan, Sweden, Ukraine, the United Kingdom of Great Britain, and Northern Ireland and the United States of America. Possible approaches to education and awareness-raising among life scientists. Working paper submitted to the Seventh Review Conference of the Biological Weapons Convention. BWC/CONF.VII/WP.20 (2011). Accessed at: http://www.unog.ch/__80256ee600585943.nsf/%28httpPages%29/f1cd974a1fde4794c125731a0037d96d?OpenDocument&ExpandSection=1#_Section1; and Revill, J., et al. Lessons learned from implementing education on dual-use in Austria, Italy, Pakistan and Sweden. Medicine, Conflict and Survival. 28(1):31–44 (2012).

8. Australia, Canada, Japan, New Zealand, Republic of Korea, and Switzerland (on behalf of the JACKSNNZ) and Kenya, Pakistan, Sweden, Ukraine, the United Kingdom of Great Britain and Northern Ireland, and the United States of America. Possible approaches to education and awareness-raising among life scientists. Working paper submitted to the Seventh Review Conference of the Biological Weapons Convention. BWC/CONF.VII/WP.20 (2011). Accessed at: http://www.unog.ch/__80256ee600585943.nsf/%28httpPages%29/f1cd974a1fde4794c125731a0037d96d?OpenDocument&ExpandSection=1#_Section1.
9. Ibid.
10. Ibid.
11. Ibid.
12. AAAS. Professional and Graduate-Level Programs on Dual-use Research and Biosecurity for Scientists Working in the Biological Sciences (2008). Accessed at: http://cstsp.aaas.org/content.html?contentid=1899.
13. NSABB. Progress Report of the NSABB Working Group on Outreach and Education. NSABB Meeting (June 23, 2011). Accessed at: http://oba.od.nih.gov/oba/biosecurity/meetings/June2011/NSABB_Outreach_Education_WG_June_23_2011.pdf.
14. Australia, Canada, Japan, New Zealand, Republic of Korea, and Switzerland (on behalf of the JACKSNNZ) and Kenya, Pakistan, Sweden, Ukraine, the United Kingdom of Great Britain and Northern Ireland, and the United States of America. Possible approaches to education and awareness-raising among life scientists. Working paper submitted to the Seventh Review Conference of the Biological Weapons Convention. BWC/CONF.VII/WP.20 (2011). Accessed at: http://www.unog.ch/__80256ee600585943.nsf/%28httpPages%29/f1cd974a1fde4794c125731a0037d96d?OpenDocument&ExpandSection=1#_Section1.
15. Ibid.
16. Ibid.
17. Ibid.
18. Revill, J., et al. Lessons learned from implementing education on dual-use in Austria, Italy, Pakistan and Sweden. Medicine, Conflict and Survival. 28(1):31–44 (2012).
19. Ibid.
20. Ibid.
21. See Majidi, V. Statement Before the Senate Committee on Homeland Security and Governmental Affairs. United States Senate (October 18, 2011). Accessed at: www.hsgac.senate.gov/download/majidi-testimony.
22. BBSRC, MRC, and Wellcome Trust. Managing risks of misuse associated with grant funding activities. A joint BBSRC, MRC and Wellcome Trust policy statement (2005). Accessed at: http://www.bbsrc.ac.uk/organisation/policies/position/public_interest/misuse_of_research_joint.pdf .
23. European Union. Getting Through Ethics Review. EU Community Research and Development Information Service (2011). Accessed at: http://cordis.europa.eu/fp7/ethics_en.html#ethics_cl.
24. Malakoff, D. U.S. Agencies to Start Screening Biomedical Proposals for Dual Use. *Science*. Vol. 336 no. 6077 (2012). Accessed at: http://www.sciencemag.org/content/336/6077/21.summary.
25. BWC. Report of The Meeting of States Parties. BWC/MSP/2005/3 (December 14, 2005). Accessed at: http://www.opbw.org/new_process/msp2005/BWC_MSP_2005_3_E.pdf.

26. Inter Academy Panel. IAP Statement on Biosecurity (November 7, 2005). Accessed at: http://www.interacademies.net/File.aspx?id=5401.
27. Biological Weapons Convention. Final Document. PART II, Final Declaration of the Second Review Conference of the Parties to the BWC. BWC/CONF.II/13/II (1986). Accessed at: http://www.opbw.org/rev_cons/2rc/docs/final_dec/2RC_final_dec_E.pdf.
28. See, for example, the flyers for Federal Bureau of Investigation Academic Biosecurity Workshops (2012). Accessed at: http://www.phri.org/news/FBI_Academic_Biosecurity_Workshop.pdf.
29. See Majidi, V. Statement Before the Senate Committee on Homeland Security and Governmental Affairs. United States Senate (October 18, 2011). Accessed at: www.hsgac.senate.gov/download/majidi-testimony.
30. Lempinen, E.W. FBI, AAAS Collaborate on Ambitious Outreach to Biotech Researchers and DIY Biologists. American Association for the Advancement of Science News (2011). Accessed at: http://www.aaas.org/news/releases/2011/0401fbi_biosecurity.shtml.
31. Mancini, G., and Revill, J. Fostering the Biosecurity Norm: Biosecurity Education for the Next Generation of Life Scientists. Working paper produced for the Landau Network-Centro Volta (2008). Accessed at: http://www.brad.ac.uk/bioethics/media/SSIS/Bioethics/docs/European_Case_study.pdf.
32. Sandia National Laboratories. 2004. Symposium Summary. International Biosecurity Symposium: Securing High Consequence Pathogens and Toxins. Sandia Report. SAND2004-2109 (June 2004). Accessed at: http://www.biosecurity.sandia.gov/sub-pages/pastConf/20032005/securing/01Int-Symposium-Summary-Final.pdf.
33. Sandia National Laboratories. *Symposium Summary.* International Biosecurity Symposium: Securing High Consequence Pathogens and Toxins. Sandia Report. SAND2004-2109 (June 2004). Accessed at: http://www.biosecurity.sandia.gov/sub-pages/pastConf/20032005/securing/01Int-Symposium-Summary-Final.pdf.
34. See, for example, the U.K. educational efforts in the chemical sector, which were abandoned due to lack of interest. Australia, Canada, Japan, New Zealand, Republic of Korea, and Switzerland (on behalf of the JACKSNNZ) and Kenya, Pakistan, Sweden, Ukraine, the United Kingdom of Great Britain and Northern Ireland, and the United States of America. Possible approaches to education and awareness-raising among life scientists. Working paper submitted to the Seventh Review Conference of the Biological Weapons Convention. BWC/CONF.VII/WP.20 (2011). Accessed at: http://www.unog.ch/__80256ee600585943.nsf/%28httpPages%29/f1cd974a1fde4794c125731a0037d96d?OpenDocument&ExpandSection=1#_Section1.
35. Australia, Canada, Japan, New Zealand, Republic of Korea, and Switzerland (on behalf of the JACKSNNZ) and Kenya, Pakistan, Sweden, Ukraine, the United Kingdom of Great Britain and Northern Ireland, and the United States of America. *Possible approaches to education and awareness-raising among life scientists.* Working paper submitted to the Seventh Review Conference of the Biological Weapons Convention. BWC/CONF.VII/WP.20 (2011). Accessed at: http://www.unog.ch/__80256ee600585943.nsf/%28httpPages%29/f1cd974a1fde4794c125731a0037d96d?OpenDocument&ExpandSection=1#_Section1.
36. Rappert, B. Introduction: Education as... In Rappert, B. (ed.). Education and Ethics in the Life Sciences—Strengthening the Prohibition of Biological Weapons. Australian National University Press. Accessed at: http://epress.anu.edu.au/education_ethics/pdf/whole_book.pdf.
37. McCartney, C., et al. 2011. Lowering the Drawbridges: Legal & Forensic Science Education for the 21st Century (2011). Accessed at: http://www.heacademy.ac.uk/assets/ps/documents/projects/law_forensics_report_2011.pdf.

38. Hoffman, J., and Thomson, G. Measuring the success of Environmental Education programmes. Canada Network for Environmental Education (2003). Accessed at: www.cpawscalgary.org/education/evaluation.
39. Revill, J. 2009. Developing Metrics and Measures for Dual-Use Education (2009). Accessed at: http://www.brad.ac.uk/bioethics/media/SSIS/Bioethics/docs/Education_metrics_and_measures_2nd_DRAFT.pdf.
40. Kelle. A. H5N1: Bungling dual-use governance. Bulletin of the Atomic Scientists (March 29, 2012). Accessed at: http://www.thebulletin.org/web-edition/columnists/alexander-kelle/h5n1-bungling-dual-use-governance.
41. Lempinen, E.W. FBI, AAAS Collaborate on Ambitious Outreach to Biotech Researchers and DIY Biologists. American Association for the Advancement of Science News (2011). Accessed at: http://www.aaas.org/news/releases/2011/0401fbi_biosecurity.shtml.
42. U.K. Parliament. Section 7—Use of Forensic Evidence in Court, Science, and Technology —Seventh Report. Select Committee on Science and Technology (2005). Accessed at: http://www.publications.parliament.uk/pa/cm200405/cmselect/cmsctech/96/9602.htm.
43. National Research Council. Strengthening Forensic Science in the United States: A Path Forward. Committee on Science, Technology, and Law (2009). Accessed at: https://www.ncjrs.gov/pdffiles1/nij/grants/228091.pdf.
44. McCartney. C., et al. 2011. Lowering the Drawbridges: Legal and Forensic Science Education for the 21st Century (2011). Accessed at: http://www.heacademy.ac.uk/assets/ps/documents/projects/law_forensics_report_2011.pdf.
45. Personal correspondence.
46. Revill, J., et al. Lessons learned from implementing education on dual-use in Austria, Italy, Pakistan and Sweden. Medicine, Conflict and Survival. 28(1):31–44 (2012).

CHAPTER 14

Monitoring and Evaluating Laboratory Biosafety Programs

Heather N. Meeks, Betiel H. Haile, Ngozi Erondu, Lisa Ferland, Meeyoung Park, Affan Shaikh, and Scott J.N. McNabb

INTRODUCTION

Biosecurity encompasses broad procedures and practices to protect humans and the environment against biological agents. It includes laboratory biosafety and biosurety as well as aspects of public health surveillance and biosurveillance of extremely dangerous pathogens and select agents that may be used intentionally (i.e., bioterrorism).[1] As stated in Chapter 1, while biosecurity pertains to practices in other settings such as agriculture and food safety, the assessment tools and indicators described in this chapter are specific to laboratory settings. Biosafety encompasses many aspects of biosecurity; here we consider it exclusively in terms of laboratory biosafety (Figure 14.1). The definition of laboratory biosafety from the World Health Organization (WHO) is "the containment principles, technologies, and practices that are implemented to prevent the unintentional exposure to pathogens and toxins or their accidental release."[2] This includes the protection of laboratory personnel and the external environment from harmful biologic agents.

The concept of laboratory biosafety encompasses the appropriate microbiological techniques, safety equipment, laboratory practices, and laboratory facility requirements necessary to protect workers and the environment. The Centers for Disease Control and Prevention (CDC) and the National Institute of Health (NIH) established four biosafety levels (BSL) to accommodate laboratory biosafety protocols for infectious agents of various classes.[3] BSL-1 and BSL-2

FIGURE 14.1 Relationships among biosecurity, laboratory biosafety, biosurety, biosurveillance, and public health surveillance

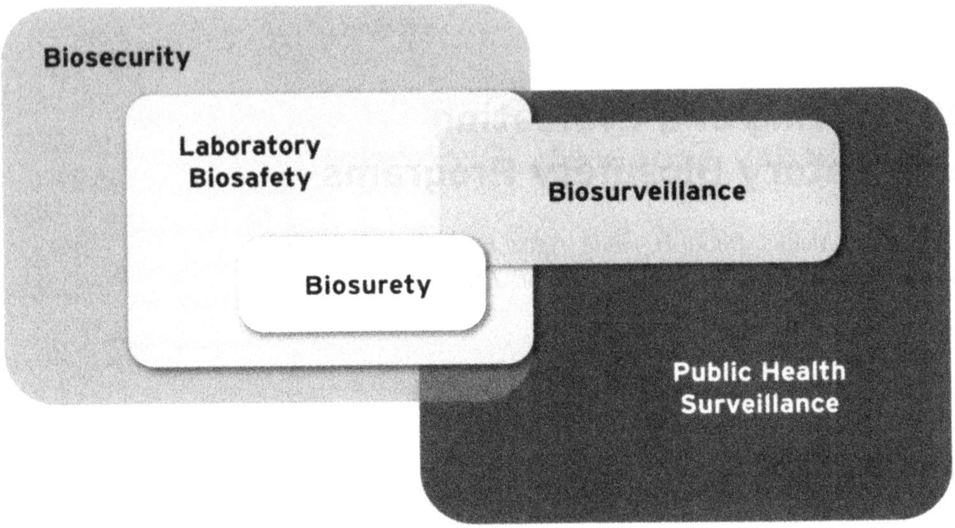

are "basic laboratories" set up to handle essential laboratory practices involving agents of moderate hazard to personnel and environment (e.g., nonpathogenic *Escherichia coli*). The "containment laboratory" is classified as BSL-3, established with all the features of BSL-1 and BSL-2 laboratories as well as additional laboratory safety policies and equipment to handle more dangerous or transmissible microorganisms (e.g., human immunodeficiency virus [HIV]). The "maximum containment laboratory" is classified as BSL-4, incorporating all features of lower BSLs and adding additional laboratory safety policies and equipment to handle the most dangerous pathogens (e.g., Ebola virus).[4] Currently, no standard, systematic, transnational, and quantitative approach is used to evaluate the factors determining laboratory biosafety program (LBP) effectiveness at different BSL levels.

This chapter describes a conceptual framework and predictive analytic tool for LBP monitoring and evaluation consisting of standardized indicators to track subsets of laboratory biosafety interventions over time and help guide future investment. The conceptual framework is flexible and can be adapted and customized for entity- or program-specific factors. Domains that make up different aspects of laboratory biosafety (e.g., biosurety, training, information security) are measured using indicators that drive the predictive features of the tool. Details related to the sensitivity and utility of these indicators enable the development of future tools and applications. This approach to monitoring and evaluation aligns to global best practices and highlights the importance of standardized assessments to identify and support specific laboratory biosafety needs of programs, countries, and regions.

NEED FOR A LABORATORY BIOSAFETY PROGRAM FRAMEWORK

Laboratory biosafety refers to the implementation of laboratory practices and procedures, specific construction features of laboratory facilities, safety equipment, and appropriate occupational health programs when working with potentially infectious microorganisms and other biological hazards (Figure 14.1).[5]

The WHO International Health Regulations (IHR) recognize the importance of laboratory safety, including Core Capacity 8—"laboratory diagnostics and confirmation capacity"—which states that laboratory services are important aspects of response, detection, and investigation with laboratory analysis of samples performed. It also stresses the need for state parties to establish mechanisms that assure reliable and timely laboratory identification of infectious agents and other hazards likely to cause public health emergencies at national and international levels.[6]

This global decree for effective and structured laboratory systems also emphasizes the importance of LBPs in protecting human health. Timely assessment and monitoring of public health program performance, specifically for LBPs, is crucial in reducing biological hazards. Unfortunately, a comprehensive framework that identifies and describes all of the components and goals of LBPs does not exist. There is no gold standard for monitoring and assessing LBPs. Further, LBP performance measures—indicators—are not consistent across existing laboratory biosafety assessments. Though some organizations (e.g., WHO and CDC) provide tools such as checklists and rubrics to gauge performance, a comprehensive source assessing the performance of LBPs at different BSLs is lacking.[7-19]

To address this gap, we developed a tool that systematically addresses subcomponents of a LBP. This tool uses specific indicators to track the successes or challenges of a LBP over time. Laboratory managers and stakeholders can use the tool to monitor key areas requiring resources, training, and personnel.

We conducted a comprehensive academic literature review through PubMed, Google Scholar, and other databases. The following key search terms were used: "monitoring and evaluation biosafety program," "laboratory indicators,"

FIGURE 14.2 Steps to develop a model for monitoring and evaluating laboratory biosafety programs

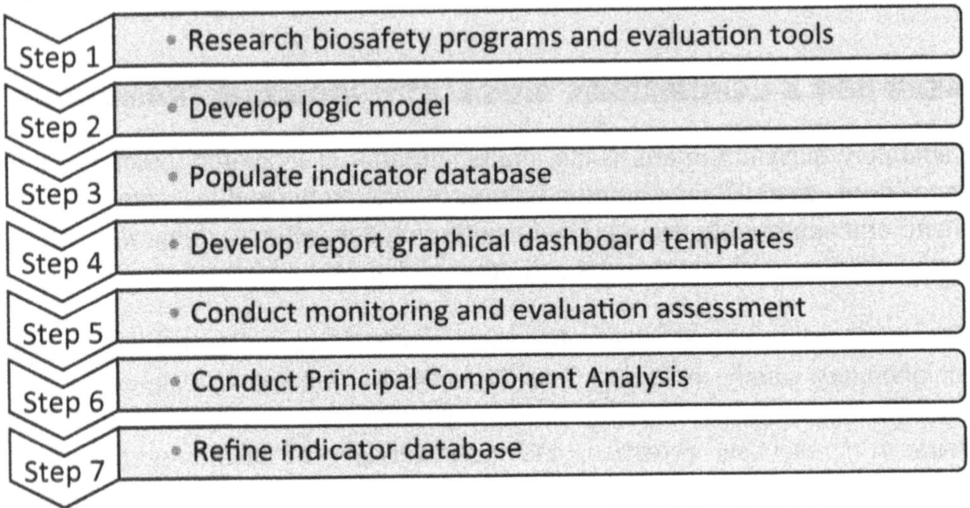

"assessing biosafety program," "biosafety outcome measurement," "program performance," "biocontainment program evaluation," "health performance assessment," "biosecurity," "biosurety," "assessing laboratory containment" and "international laboratory goals." Literature gathered included laboratory biosafety manuals that discuss configurations for biosafety programs and laboratory guidelines. Terminology, program components, and performance measures differ greatly across laboratory manuals depending on the organization and laboratory facility. This highlights the difficulty in determining best practices for laboratory biosafety strategic planning and presents a barrier to monitoring and evaluating LBPs for overall program improvement.

The Logic Model as a Conceptual Framework for LBPs

To create a standard framework for assessing a LBP, we concatenated the inputs and activities of a LBP and aligned the envisioned outcomes within each domain. The logic model is a graphical representation of a system designed to identify important elements and their relationships within that system (Table 14.1).[20] When properly constructed, the logic model illustrates the underlying hypotheses that selected intervention(s) will provide to the identified problem with observable and measureable outcomes.

Logic models have been used for program planning, management, and evaluation.[21,22] Many professional domains—from public health to criminal justice—have used logic models to understand program dynamics. In this chapter, the

TABLE 14.1 Description of laboratory biosafety logic model elements

Inputs*	Activities†	Outputs§	Outcomes¶	
• Time • Money • Partners • Personnel	• Trainings • Develop documents • Perform background checks	• Trained employees • Standard operating manuals • Completed personnel security checks	Short Term	Describes immediate and measurable changes in several aspects of the institution as a direct result of the program
			Intermediate	Occurs after the program has been implemented for a longer period of time and describes changes in institutional performance and practices
			Long Term	Describes expected lasting impact(s) of the sustained program; should be the ultimate goal of the program

Adapted from http://www.uiweb.uidaho.edu/extension/LogicModel.pdf

*What we invest
†What we do
§What we produce (from activities)
¶Results of the program

logic model was used as a conceptual framework for a LBP and graphically represents program processes and workflow. The purposes of the LBP logic model were to:

- Build a common understanding of the overall structure and function of a LBP
- Identify LBP activities and ensure that they correspond to intended outcomes
- Determine a set of key performance measurement indicators
- Reveal where processes in the LBP could break down

For the purpose of this chapter, *domains* are defined as the specific focus of the sublogic models (e.g., procedural security), and *elements* are the vertical categories or columns within the logic model (e.g., inputs, outcomes). Within the logic model, *components* refer to items populating the elements. When discussed later in the chapter in terms of principal component analysis (PCA), it has an alternate meaning, referring to the artificial variables used in the analysis that may assess more than one component of the logic model.

Figures 14.3 to 14.8 display the primary logic model for laboratory biosafety and the five sub-logic models containing measures for evaluation and assessment for each domain of biosafety.

Applying the Principal Component Analysis to Laboratory Biosafety

The lack of standardized indicators to assess laboratory biosafety makes it a challenge to objectively measure both LBP performance and attainment of program goals. Although various methods seek to define laboratory biosafety and goals, few incorporate a quantitative means to measure progress, evaluate return on investment (ROI), and predict a likely trajectory of progress. Proposed as a multitiered approach to minimize the burden of the assessment while providing an end product useful to stakeholders, PCA is a useful method to address the above objectives.

MAJOR CONCEPTS AND DEFINITIONS

There are many organizations within public health, biosecurity, and government that utilize the same terms but define them according to their own perceptions or purposes. WHO defines laboratory biosecurity as "the protection, control and accountability for valuable biological materials (VBM) within laboratories in order to prevent their unauthorized access, loss, theft, misuse, diversion, or intentional release." The key difference between laboratory biosafety and laboratory biosecurity, as defined by WHO, is the larger focus on bioterrorism that laboratory biosecurity aims to prevent.[4] In contrast, the Food and Agriculture Organization (FAO) of the United Nations defines biosecurity as "a holistic concept of direct relevance to the sustainability of agriculture and wide-ranging aspects of public health and protection of the environment, including biological diversity." This includes the risks related to food safety, animal life and health, plant life and health, and associated environmental risks.[23]

Definitions

For the purposes of this chapter, we define the following terms with the understanding that there are many diverging opinions and definitions about the scope and breadth of each. Terms related to biosafety here align to concepts presented in Chapter 1 or are otherwise adapted with explanations as to the discrepancy in the definition. Definitions were adapted through the lens of public health.

(Text continued on page 237)

FIGURE 14.3 Primary logic model for laboratory biosafety

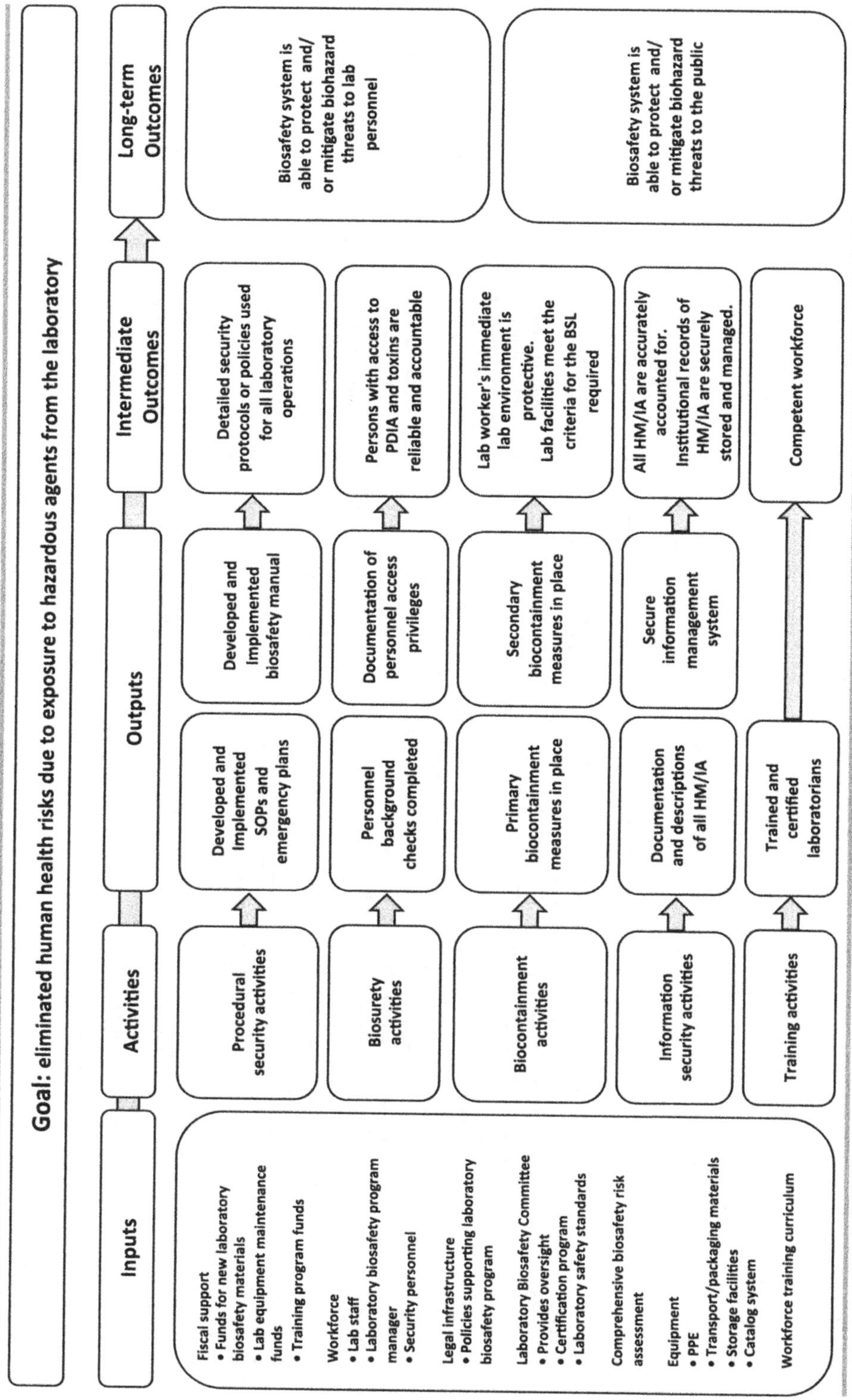

FIGURE 14.4 Procedural security sublogic model for laboratory biosafety

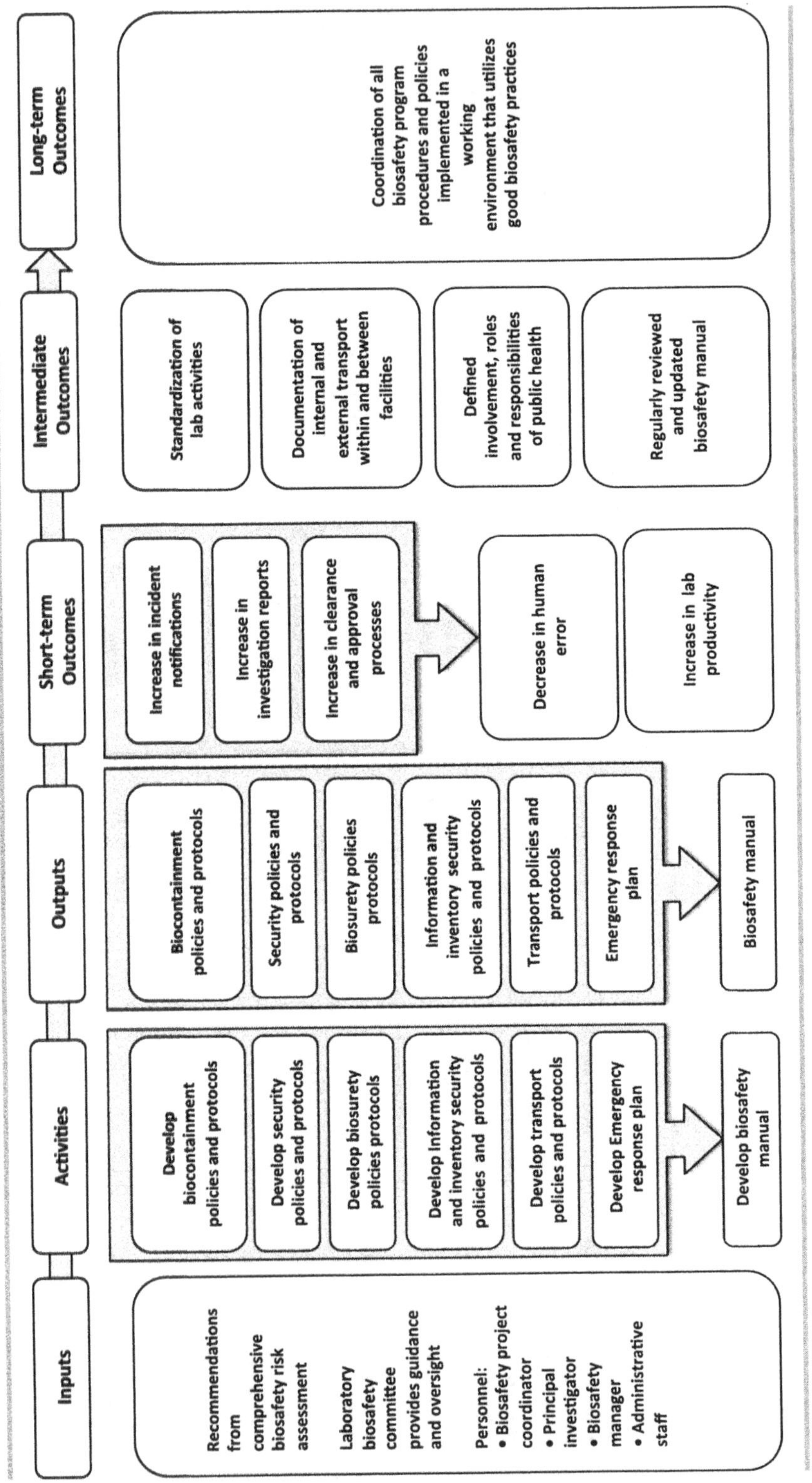

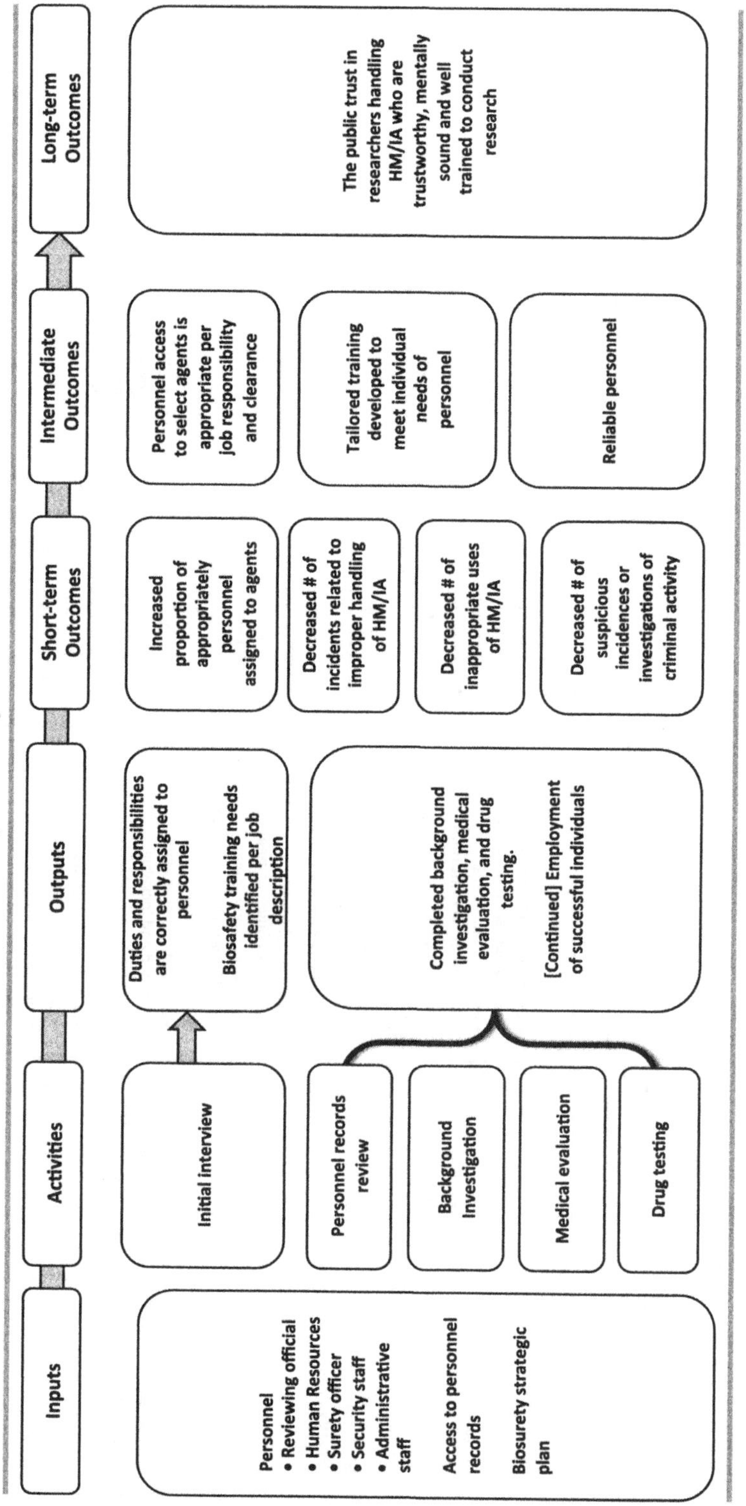

FIGURE 14.5 Biosurety sublogic model for laboratory biosafety

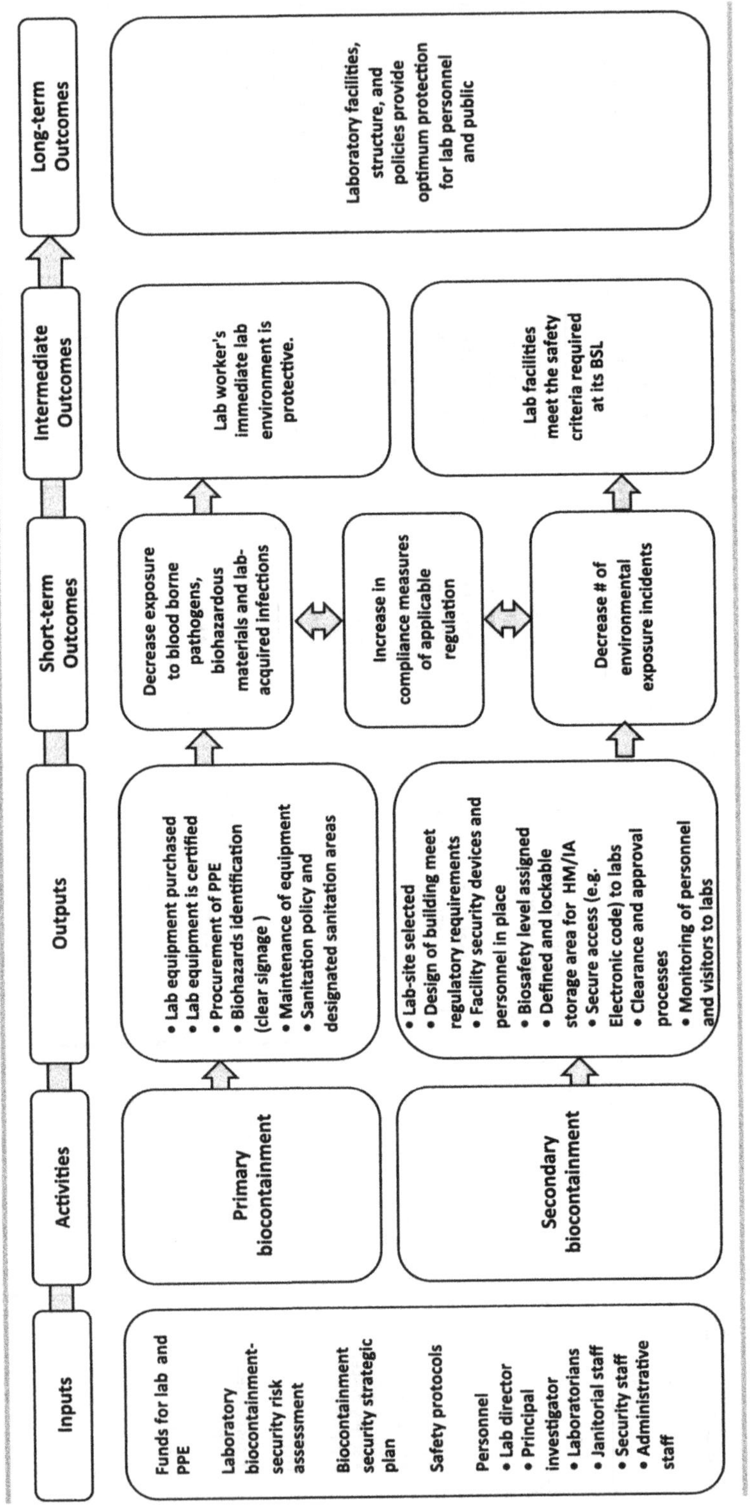

FIGURE 14.6 Physical security sublogic model for laboratory biosafety

FIGURE 14.7 Information security sublogic model for laboratory biosafety

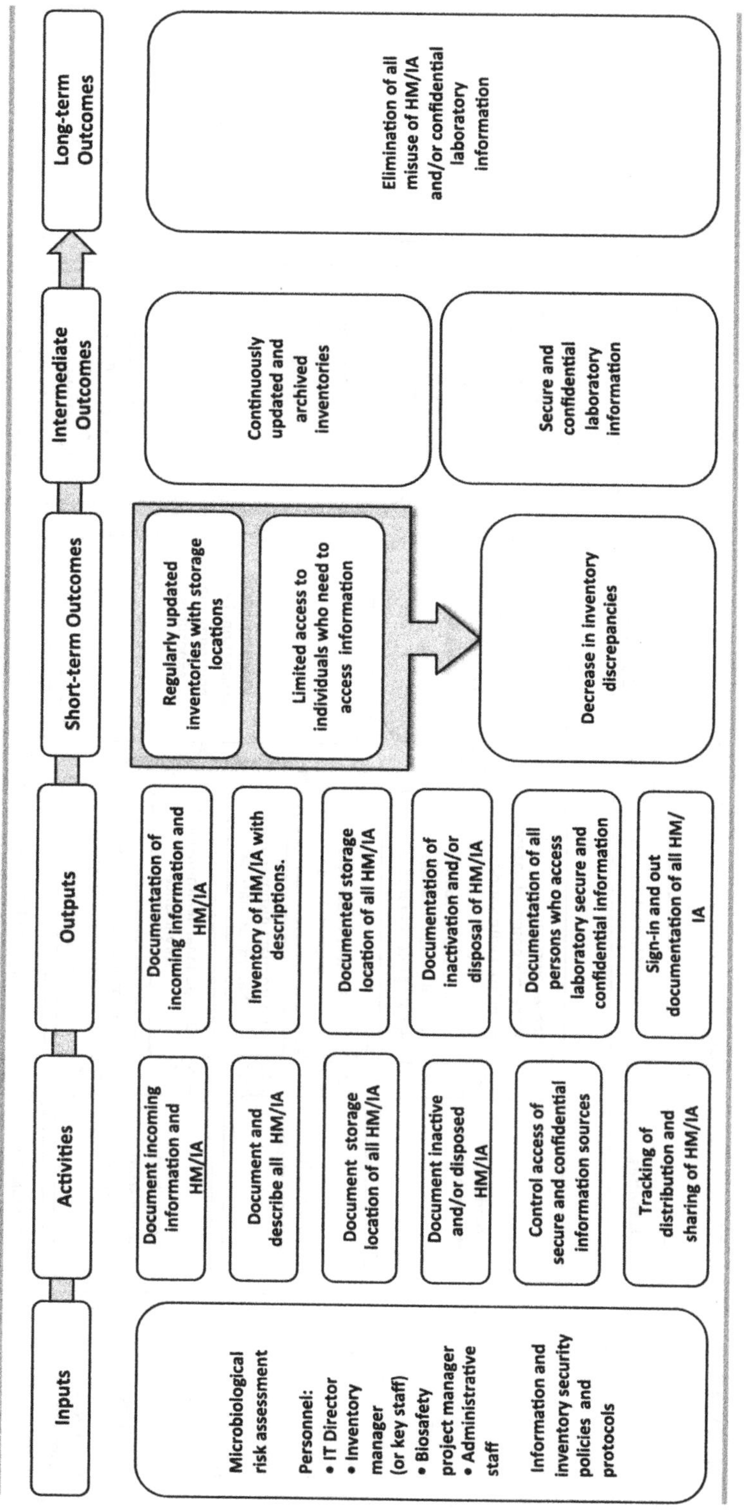

FIGURE 14.8 Training sublogic model for laboratory biosafety

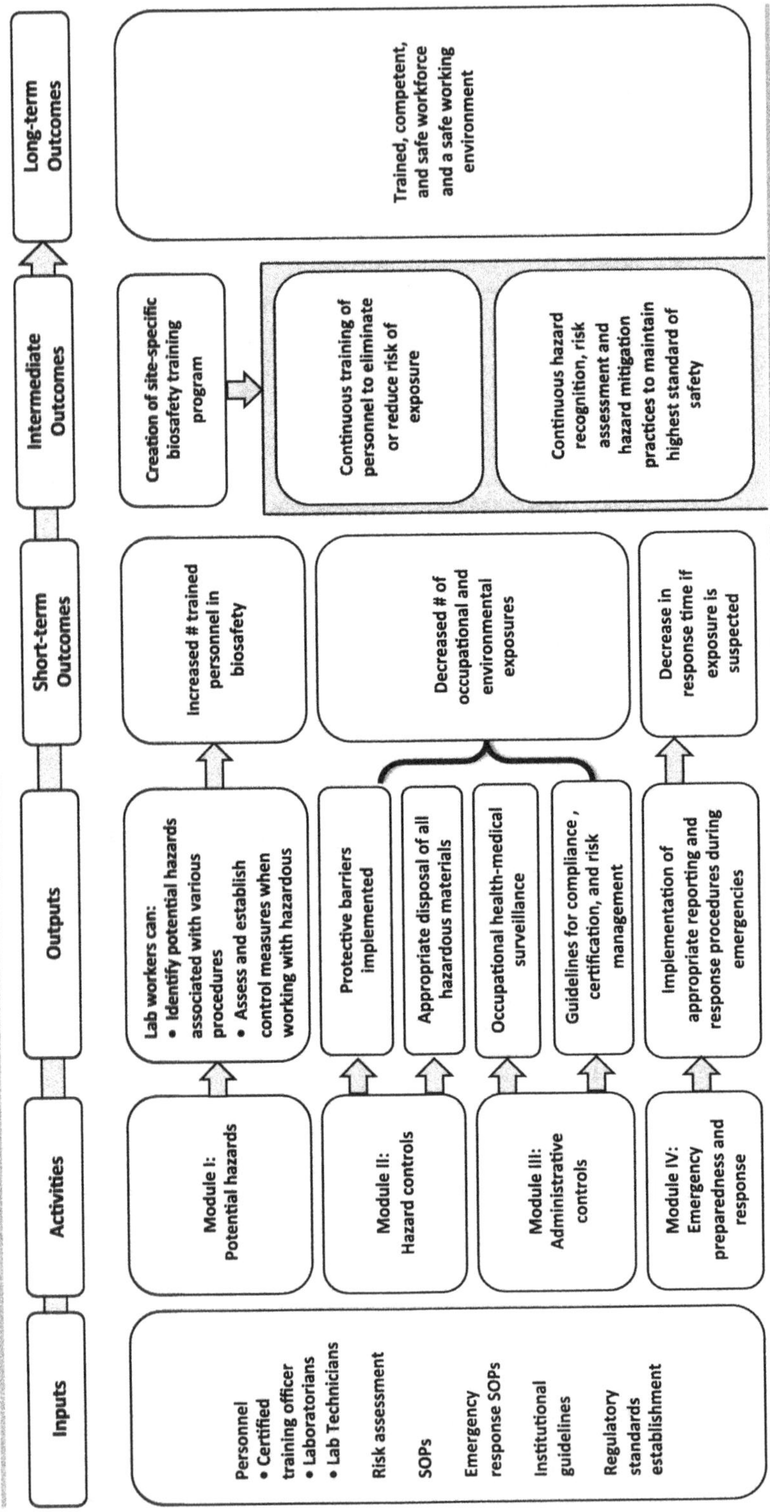

Physical Security

The safety practices and procedures used to prevent unintended infection employ the physical measures of biocontainment (primary and secondary) and isolation to protect laboratory workers and the environment from the agents used in a laboratory. Examples of biocontainment barriers in a facility include contained dressing and shower rooms, sealed service penetrations, specialized doors, entry and exit avenues to prevent cross-contamination, specialized air handling systems for containment control, personal protective equipment, and biosafety cabinets[24] (Chapter 1).

Biosecurity

Biosecurity is the protection, control, and accountability for valuable biological materials within laboratories in order to prevent unauthorized access, loss, theft, misuse, diversion, or intentional release.[2]

Biosurety

Biosurety encompasses the principles and procedures as they relate to the security of laboratory personnel (Chapter 1). Biosurety generally applies only to BSL-3 and BSL-4 facilities and is a practice found within laboratories run by the US Department of Defense, such as at the United States Army Medical Research Institute for Infectious Diseases. Biosurety policies encompass aspects of biosafety, personnel responsibility, and agent accountability.[25]

Biosurveillance

Biosurveillance is the process of active data gathering with appropriate analysis and interpretation of biosphere data that might relate to disease activity and threats to human or animal health—whether infectious, toxic, metabolic, or otherwise and regardless of intentional or natural origin—in order to achieve early warning of health threats, early detection of health events, and overall situational awareness of disease activity.[2]

Information Security

Information security encompasses policies established to control access to sensitive information related to security plans, access control codes, and inventories. Information security is critical to establish inventory security and other management procedures used to track inventory, storage, use, transfer, and destruction of biological materials. While many laboratories utilize integrated electronic systems to manage information security, low-resource laboratories may use paper records or other low-tech means as a management system.

Laboratory Biosafety

Laboratory biosafety refers to a set of principles and practices that dictate the safe handling and containment of potentially harmful biological agents for the sake of preventing infection of laboratory workers and the public (Chapter 1).

Primary Biocontainment

Primary biocontainment is the protection of personnel and the immediate laboratory environment from exposure to infectious agents.[26]

Procedural Security

Procedural security refers to security protocols or plans for assessing risks and safety considerations of activities in the laboratory environment, including quality management (laboratory accreditation) and transport security.[27]

Public Health Surveillance

Public health surveillance is the ongoing, systematic collection, analysis, interpretation, and dissemination of data regarding a health-related event for use in public health actions to reduce morbidity and mortality and to improve health.[28]

Secondary Biocontainment

Security biocontainment is the protection of the environment external to the laboratory from exposure to infectious materials, as provided by facility design and operational practices.[26]

Training

Training refers to programs to inform and educate individuals regarding their responsibilities within the laboratory and the institution.[29]

CONCEPTUAL FRAMEWORK

The laboratory biosafety program conceptual framework defines the elements necessary for LBPs to operate.

Creating the LBP Logic Models: Goals and Outcomes

Following an extensive literature review and initial research, we developed the primary LBP logic model and identified domains for each sublogic model. Elements for the main logic model included inputs, activities, outputs, and intermediate- and long-term outcomes. Industry terms and standard concepts were used to supplement critical components of each domain. We used a "right-to-left" approach to design these logic models, starting with an expected result and mapping the specific processes needed to reach the result in order to portray relationships between components. Consequently, goals and stepwise outcomes were the first elements developed in this model. The goals and outcomes were derived from the definitions used earlier in this book as well as from information in the WHO Biorisk Management Laboratory Biosafety Guidance[2] and the CDC Biosafety in Microbiological and Biomedical Laboratories (BMBL).[30] The main model outcomes of a successful LBP were divided into intermediate goals, or expected outcomes, assessed at some discretionary review period or two to five years after the implementation of a LBP, and long-term outcomes, which in this case are the expected outcomes after five years.

The intermediate outcomes can be measured by a specific set of indicators that correspond to events of an activity (found in each sublogic model). For the main logic model, *long-term outcome* is synonymous to program impact. This impact may not be revealed in an evaluation of the individual program but may be evident from future nationwide data showing changes in laboratory-related incidents or health outcomes of a specified region or population.

Creating the LBP Logic Model: Activities and Outputs

Next we reviewed ten recent laboratory manuals from universities, international institutions, nongovernmental organizations, and governmental organizations,[4,23,30-37] each of which developed its own approach to achieving laboratory biosafety outcomes. From these manuals, concepts within the five domains were extracted to illustrate the core functions of a LBP: procedural biosafety, biosurety, physical security, information security, and training. Language for these LBP domains was harmonized to consolidate related sets of activities or outcomes identified in manuals as steps for increasing laboratory biosafety compliance. These domains classify and organize specific intervention activities hypothesized to improve LBP outcomes. Though the activities were arranged sequentially by stages of implementation, in practice several activities might be performed simultaneously. Each domain was further delineated into a separate sublogic model describing the steps necessary to execute the activity. The outputs were derived from the literature documenting each biosafety activity. The output is a tangible product of an activity or set of activities. Outputs were determined to be essential to demonstrating the efficacy of a LBP.

Creating the LBP Logic Model: Inputs

The last step in developing the main logic model was identifying inputs. Inputs are the primary resources necessary to launch a LBP. These components should be established prior to undertaking a program activity. Each program activity was reviewed to ascertain the human and monetary resources, policies, procedures, and organizational oversight and leadership needed to successfully execute the activity. For instance, a major input identified in nearly all laboratory manuals was a biosafety risk assessment. CDC describes the risk assessment as a process used to identify the hazardous characteristics of a known or potentially infectious agent or material. Furthermore, CDC directs laboratory management and principal investigators to use risk assessment information to alert and direct staff, control hazards with appropriate containment measures, and protect persons directly and indirectly associated with the laboratory.[30]

Sublogic models

The elements of the sublogic models were developed using a bidirectional approach—from "left to right" and "right to left." This was feasible since the activities were identified at a higher level.

Outcomes were conceived by program objectives found in the laboratory biosafety literature review. Outcomes were defined such that long-term outcomes are the intended impact of the successful activity; intermediate outcomes are the program changes as a result of an effective activity; and short-term outcomes, an element unique to the sublogic models, are measurable changes that are an expected and almost immediate result of a listed activity.

Activities in the sublogic models are more specific and represent a stepwise approach to increasing laboratory biosafety. For example, the activities in the biosurety logic model include five steps to establish biosurety according to accepted guidelines. The initial interview must be performed before any personnel review or background investigation is initiated. Upon passing these steps, the applicant may be requested to undergo a medical evaluation and a drug test if warranted by the job title.

Outputs in the sublogic models are specific results of a related activity. These products, policies, protocols, and skills were acquired from the reviewed laboratory manuals (which usually included some type of training manual).

Inputs were determined under the assumption that a laboratory biosafety risk assessment had been performed and completed. The information and recommendations provided from this assessment would identify which resources are needed to conduct a set of activities within each domain. Conducting the assessment will help decision and policy makers to effectively disseminate resources and fill gaps in fundamental program inputs to facilitate the execution of needed activities.

LABORATORY BIOSAFETY INDICATORS

The LBP indicator database measures performance and identifies gaps, opportunities, and challenges for strategic planning. The indicators drill into a program area to track progress or identify obstacles in meeting facilitywide goals. Accurate data collection using indicators provides objective information to facilitate the improvement of LBPs and advocate for resources. Key stakeholders include local, national, regional, and international laboratories.

All indicators were grouped into a specific domain in the five LBP sublogic models. Indicators were then mapped to the sublogic models' elements: inputs, activities, outputs, and short-term outcomes. The indicator database contains the following:

- Indicator description: a concise explanation of the indicator
- Biological safety level (BSL): a drop-down list of the biological safety levels the indicator measures in BSL 1 to 4
- Program logic model map: a drop-down list of logic model items linked with each indicator
- Definition: detailed definition of the indicator
- Numerator data source: the source where the numerator data can be found
- Denominator data source: the source where the denominator data can be found
- Source (e.g., WHO, CDC): the original source of development of the indicator
- Geographic scope (i.e., local, national, regional, international): a drop-down list of the geographical area where the indicator can be applied

- Answer type: the expected answer type for the indicator
- Target: the target the program sets for that indicator
- Reference: the reference where the indicator can be found

Indicators were collected from relevant articles and laboratory safety manuals of the WHO, CDC, Association of Public Health Laboratories, and other indicator resources. Most of the indicators were developed from information gathered from the articles and manuals, and they measure components (e.g., a specific outcome) where existing indicators did not.

ANTICIPATED BENEFITS OF THE INDICATOR DATABASE

This indicator database contains information for developing self- or external assessments of LBPs. The initial assessment establishes baseline data for comparison of subsequent years' data to identify areas for improvement or progress. This can provide useful information for decision makers in prioritizing more efficient use of resources and targeted interventions. The indicator database will help laboratory biosafety program mangers or stakeholders address the following questions:

1. How to group a program's activities in a clear and logical manner
2. How to choose an outcome measure, given the data available, to track the level of performance for each program activity
3. How the program performs from year to year
4. Which areas of the program are performing up to target goals and which ones are not

Though indicators were developed for the domains of each sublogic model, indicators selected for the final evaluation and assessment tool will need to be refined by using additional analytical approaches.

ANALYTIC PROCESSES TO REFINE INDICATORS

Rationale for a Multitiered Approach

A robust analytical framework provides the ability to assess LBPs, to predict the likely program trajectory based upon the current state, and to evaluate the impacts of alternative courses of action (e.g., targeted investment in discrete program elements). The most useful framework will incorporate an ordered and systematic approach involving multiple layers of analysis so that a useful picture of the data emerges. The approach described here leverages commonly used statistical analyses, including principal components and regression-based analyses, in a stepwise fashion in order to address our objectives.

PCA is a data reduction technique to decrease the number of variables (i.e., indicators) in a dataset, provided there are redundancies in the variables. PCA extracts important information from the dataset to reveal hidden dynamics or patterns not evident by evaluating values for each indicator independently. It identifies areas of redundancy in a given indicator set and collapses the indicators into a smaller number of artificial variables, called principal components, that account for a substantial amount of the total variance (equal to the number of indicators analyzed) in the dataset.[38,39,40] It evaluates large and complex datasets by finding correlations among variables and creating a reduced set of variables that nonetheless represents most of the information (i.e., variance) in the original data.[39]

In addition, PCA is useful for identifying emerging patterns and relationships that are difficult to elucidate using purely regression-based analyses.[41] Statistical programs can also approximate weights for the variables within each principal component that will allow "scores" to be computed for individual LBPs. The principal components selected as a result of conducting the analysis can subsequently be used as predictor variables for regression-based or other statistical analyses in order to evaluate strength of correlations, if any, between the dependent (or "criterion") and independent (or "predictor") variables. Results can be presented graphically in a manner that is easily interpreted and user-friendly without compromising statistical rigor.

Precedence

PCA may be the most widely used multivariate (i.e., involving more than a single variable) statistical technique, and it is applied to data analysis in virtually every scientific discipline. PCA is a common means of reducing dimensionality in data prior to other forms of analysis, as it a priori identifies and eliminates redundant data collection, reduces the analytical burden related to performing numerous regression analyses, and elucidates the cumulative impact of indicators that, individually, may appear to bear little relationship to outcomes.[41,42] Because the technique is computationally laborious, PCA came into popular use following the dawn of the computer era and has since become a standard form of analysis in fields such as evolutionary biology,[43] ecology,[44] health and nutrition,[45] economics,[46] and the social sciences.[47] PCA can be used as a stand-alone analysis, but greater refinement of results will be attained by performing additional analyses (e.g., multiple regression) that can assess the strength and nature of relationships between predictor and criterion variables and inform predictive modeling.[40]

Eliminating Redundancy

The concept of variable redundancy is illustrated by the example presented in Table 14.2 (information is fictitious). PCA is not limited to evaluating survey data and may be used on many different types of inputs, although it may not perform as desired on binary (i.e., "1-0" or "yes-no") data or on a dataset for which some proportion is composed of binary data. Other options, including extensions of PCA, are available for different data types.[40]

PCA is performed in a series of steps, which are described in more detail below. The analysis described results in an orthogonal solution (i.e., retained components will be uncorrelated).

TABLE 14.2 An illustration of laboratory biosafety variable (i.e., indicator) redundancy for principal component analysis

Indicators
1. Funding levels for new biosafety materials are appropriate
2. Laboratory equipment maintenance funds are sufficient
3. Funding is adequate to ensure that personnel receive proper training
4. Biosafety guidelines or recommendations provide for appropriate personnel protection
5. Regulations for physical protection of facilities are adequate to prevent security breaches
6. Regulations for agent transport are sufficient to provide for appropriate personnel protection
7. Information security procedures are adequate to prevent inappropriate dissemination of data

Indicators 1 – 3 appear to assess issues related to adequate funding, whereas indicators 4 – 7 appear to evaluate guidelines or regulatory standards.

Generate Data on the Selected Indicators

Better results are obtained when decisions are made a priori about the constructs to be measured. Analysis of laboratory biosafety practices should be informed by first conducting a thorough review of current methods of evaluation, using the results of the review to derive the putative overarching domains that define and measure biosafety, and developing indicators to measure each domain. The framework, including the logic model and indicator database, described in this chapter represents an important precursor to the backend statistical analyses recommended here.

PCA is a large-sample procedure; therefore, the general standard for adequate sampling is given as five times the number of indicators or 100 samples, whichever is greater. For example, suppose that 50 indicators are chosen to evaluate laboratory biosafety programs. Five times the number of indicators is equal to 250, and thus 250 samples, for which data are complete, should be represented in the final analysis.

Generate a Correlation or Variance-Covariance Matrix

PCA can be performed on several types of data, but, because the indicators measuring LBPs are expected to have different units or scales for measurement, a correlation matrix may be most appropriate (Table 14.3). The values in the table are "correlation coefficients" and may be calculated using different methods. The example matrix reveals high levels of correlation among indicators 1 to 3 ($\geq$0.78) and 4 to 7 ($\geq$0.76) and low levels of correlation ($\leq$0.05) between the two indicator sets. The value of using a matrix as opposed to other data formats is that it provides a sense of how the data are clustered and, likewise, may provide a basis for reevaluating and removing some indicators as the model is refined.

Extract the Principal Components and Determine the Number to Retain

The number of indicators used for the analysis dictates the number of principal components extracted. The first component will account for the largest amount of variance in the dataset, and successive components will account for progressively smaller amounts. Only the first few components will typically be retained for further analysis, because the variance attributed to later components is often trivial (Table 14.4). Statistical programs can be used to analyze the correlation matrix and produce a table of "eigenvalues," each of which represents the variance accounted for by a given component.

TABLE 14.3 Principal component analysis correlation matrix for laboratory biosafety

	Pearson's Correlation Coefficient Values* (r)						
Indicators	**1**	**2**	**3**	**4**	**5**	**6**	**7**
1	1.00						
2	**0.78**	1.00					
3	**0.89**	**0.87**	1.00				
4	0.04	0.03	0.02	1.00			
5	0.03	0.01	0.01	**0.76**	1.00		
6	0.01	0.02	0.05	**0.85**	**0.88**	1.00	
7	0.02	0.04	0.00	**0.79**	**0.82**	**0.92**	1.00

***All values are fictitious**
Correlations between two indicators are denoted at the intersection of the row and column for a given indicator set. Values of ≥ 0.70 indicate strong correlation.

The next step is to interpret the eigenvalues in order to determine which of the principal components should be retained for analysis. Several criteria can be applied to assess whether components are meaningful.[48] The criteria may be applied independently, but, in order to avoid arbitrarily dismissing components that may actually be significant, a better approach is to apply each of them sequentially.

Rotate the Factor Pattern Matrix to Improve Interpretation

Once the principal components are extracted, an "unrotated" factor pattern matrix can be produced in which the rows of the matrix contain the indicators and the columns contain the factors. The factor pattern matrix can be evaluated to determine the overarching domains being represented by the individual components; however, interpretation of an unrotated matrix is often challenging. A linear transformation can be performed to simplify interpretation, as each indicator will tend to be associated with one or a small number of components. The components, in turn, should represent a limited number of indicators. The end goal of rotation is to divide the indicators into distinct sets so that the domains being evaluated by each component are easily discernible. The component may then be labeled according to the domain it seems

TABLE 14.4 Eigenvalue analysis for laboratory biosafety principal component analysis

	Eigenvalues of the Correlation Matrix: Total = 7			
	1	**2**	**3**	**4**
Eigenvalue*	2.7034	1.9735	0.9877	0.5600
Proportion†	0.3677	0.3323	0.1279	0.0632
Cumulative§	0.3677	0.700	0.8279	0.8911
	5	**6**	**7**	
Eigenvalue*	0.2816	0.2752	0.2186	
Proportion†	0.0479	0.0385	0.0225	
Cumulative§	0.9390	0.9775	1.000	

Eigenvalues represent the total amount of variation in the dataset explained by the components (labeled 1 – 7) extracted from the correlation matrix. If the eigenvalue-one, scree test, proportion of variance, and interpretability criteria are applied, the first three principal components will be retained.
† Proportion denotes the proportion of total variance for which each component accounts.
§Cumulative value reflects the increasing proportion of variance accounted for as the components are summed sequentially.

to measure. The final solution should be evaluated to determine that it meets baseline interpretability criteria:

- Three indicators, at minimum, have high loadings (absolute value >0.40) on each of the retained components.
- The indicators that load on a given component are conceptually similar.
- The indicators that load on different components are conceptually dissimilar.
- Indicators have high loadings on only a single component and near-zero loadings on all other components. Likewise, components should have high loadings for some indicators but near-zero loadings for all others.

If these criteria are used to evaluate the factor pattern matrix (Table 14.5), components 1 and 2 will be retained. Component 1 seems to represent the overarching domain of adequate funding, whereas component 2 apparently

TABLE 14.5 Rotated factor pattern matrix for laboratory biosafety principal component analysis

	Factor 1*	Factor 2*	Factor 3
Indicator 1	0.00	0.91	0.03
Indicator 2	0.03	0.71	0.11
Indicator 3	0.07	0.85	-0.09
Indicator 4	0.85	0.09	0.10
Indicator 5	0.90	0.08	-0.07
Indicator 6	0.81	0.15	0.01
Indicator 7	0.10	-0.02	0.79

If the interpretability criteria are applied, factors (i.e., principal components) 1 and 2 are retained. Only one indicator loads significantly (as indicated by absolute value ≥0.40) on factor 3, thus it will be omitted.

applies to appropriateness of regulatory standards and guidelines. The components can be named accordingly.

Create Factor Scores on Individual Samples

Scores can be assigned to individual samples in order to determine status with respect to retained components (Table 14.6). For example, an LBP might be described by two components, one of which is monetary investment and the other is supporting regulatory standards and guidelines. The effectiveness of LBPs in particular laboratories, regions, or countries can be evaluated based on the calculated scores, and comparison of scores can reveal the standing of individual programs relative to others. The scores can be used also as predictor variables in regression-based analyses, such that the likely program trajectory with and without intervention may be evaluated.

Evaluate the Quality of the Model

For the model to be useful, it should be agnostic to region, provide predictive power, and invoke the principle of maximum parsimony. As such, assessing its generalization capacity by estimating the value of new observations represents a primary goal.[32] Resampling procedures such as the bootstrap (sample with replacement) method can be used to estimate probability distributions for the larger population from which the discrete data are gathered and to establish confidence intervals. Additionally, cross-validation methods can inform estimations of predictive error.

Following the procedures denoted above can also aid in evaluating the appropriate number of components to retain. The objective is to develop a model

TABLE 14.6 Indicators, calculated weights, and descriptions for analysis of laboratory biosafety principal component analysis

Indicator	Weight*	Description
Component 1: Monetary Investment		
Funding for laboratory biosafety materials	0.15	Funding levels for new biosafety materials are available and adequate
Equipment maintenance funds	0.40	Laboratory equipment maintenance funds are available and adequate
Personnel training funds	0.45	Funding levels for training of new and existing laboratory personnel is available and adequate
Component 2: Appropriateness of Regulatory Infrastructure		
Guidelines for personnel protection	0.4	Biosafety guidelines and/or recommendations support appropriate levels of personnel protection
Regulations for facility protection	0.4	Regulations for physical protection of facilities are adequate to prevent security breaches
Regulations for agent transport	0.2	Regulations for agent transport are sufficient to provide for appropriate personnel protection

Weights of the indicators were calculated during PCA to assign scores to individual measures. The weights provide a basis to identify specific targets for intervention and estimating levels of investment most likely to yield positive returns.

that is sufficiently explanatory while minimizing the amount of information required. It is important to remember that adding more components does not always increase the quality of prediction.

Regression-based Analysis

PCA is frequently used as a precursor to other statistical techniques, especially those sensitive to muticollinearities (i.e., where two or more predictor variables are strongly correlated). For the LBP logic model, using the principal components as predictor variables for subsequent regression-based analysis may help to tease out important relationships between the components (e.g., monetary investment and availability/implementation of appropriate regulatory standards) and a particular criterion variable of interest (e.g., reduction in number of biohazard incidents). Regression-based analysis can assist in determining which variables represent useful predictors and, in turn, provide a starting point for refining the number and types of data that are gathered to assess a particular criterion variable. Classical multiple regression, or extensions of the technique tailored to PCA, may be applied for such evaluations.

An example of presentation of the predictive ability of PCA is provided in Figure 14.9. The x-axis measures laboratory biosafety program regulation infrastructure from "poor/minimal" to "above average," and the y-axis measures monetary investment in laboratory biosafety programs from "poor/minimal" to "above average." Current and historical data for each region (denoted by circles) will determine both placement on the graph and predicted program trajectory (denoted by vectors associated with the circles). The graph can be divided into quadrants that indicate where a particular region resides with respect to the "optimum" condition (upper right quadrant), which is interpreted as zero or near-zero biohazard incidents for a given time period, and to identify the intervention strategies that will have the greatest impact. For example, region A might benefit from improved regulatory infrastructure, whereas region E could benefit from both increased monetary investment and improved regulatory infrastructure.

FIGURE 14.9 Potential use of principal component analysis results for laboratory biosafety programs

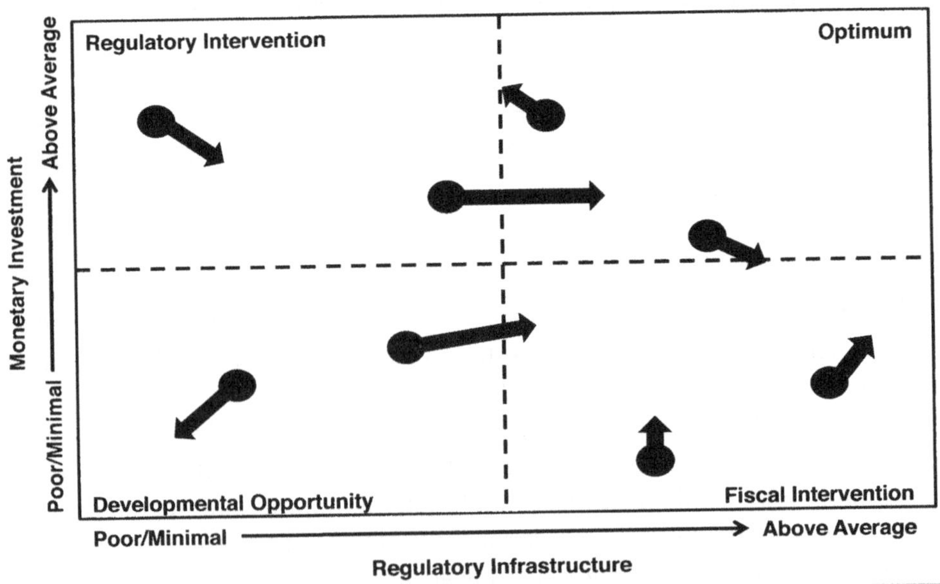

Presentation of Health Information on the Front End

The ultimate purpose of evaluating LBPs using meticulous and standardized methods is to provide stakeholders with the health information to efficiently improve quality. We developed the conceptual framework around a logical model that describes the LBP. Indicators were rigorously selected to measure the status of necessary inputs and targeted outputs of the logic model. Ultimately, a set of indicators will be refined using PCA to eliminate those found to have trivial impact in terms of the biosafety program evaluation. Through these steps, a comprehensive yet parsimonious analytic tool will be conceived.

Effective communication of analytical outputs (the front end) is also critical to aid decision makers tasked with developing intervention strategies designed to improve laboratory biosafety practices and reduce the number of biohazard incidents. Examples of results from analyses can be graphically characterized (Figures 14.10 and 14.11). If the primary goal is to identify specific targets for intervention, a data table (Table 14.7) that contains information on the indicators and their respective weights will be useful.

FIGURE 14.10 Simplified example of the measured progress versus a set of activities to reach compliance for each laboratory biosafety domain

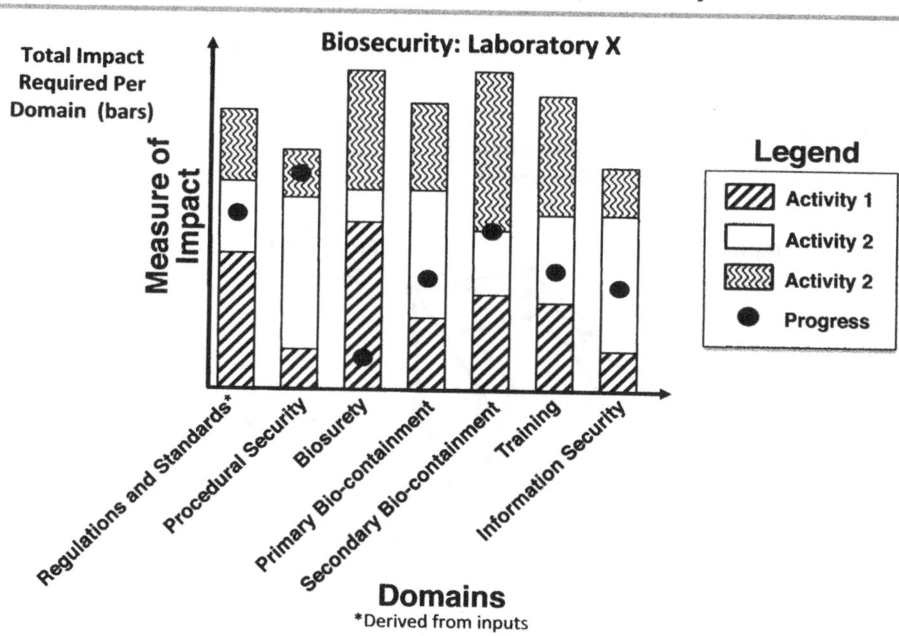

The bars in Figure 14.10 show the total estimated amount of impact for each activity within a domain. The cumulative amounts of impact for the set of activities in the element are also displayed. If a laboratory is found to have a low level of development in a domain, such as for physical security in Figure 14.10, then it is necessary to complete all listed activities to bring that domain to full compliance. For domains with high level of development, such as with procedural security or biosurety, little to no work may be necessary to meet relevant standards for that domain.

Execution of certain activities is often contingent upon the completion of a prior one. The report generated by the toolset grants the ability to prioritize and select a set of activities based on fiscal and resource capacity versus expected impact. Impact is estimated for individual and sets of activities along with estimations of cost to provide financial and business justification for implementing those activities. In the example shown in Table 14.7, the cells highlighted in the upper segment indicate which activities have yet to be conducted. The lower segment calculates the impact versus cost for each activity, with highlighted cells again corresponding to missing activities. Assessing which set of activities provides the greatest ROI allow decision makers to allocate resources accordingly. Since activities in this model are designed to be implemented

FIGURE 14.11 Spider graph showing current, expected, and actual outcomes or measures of progress in various laboratory biosafety domains

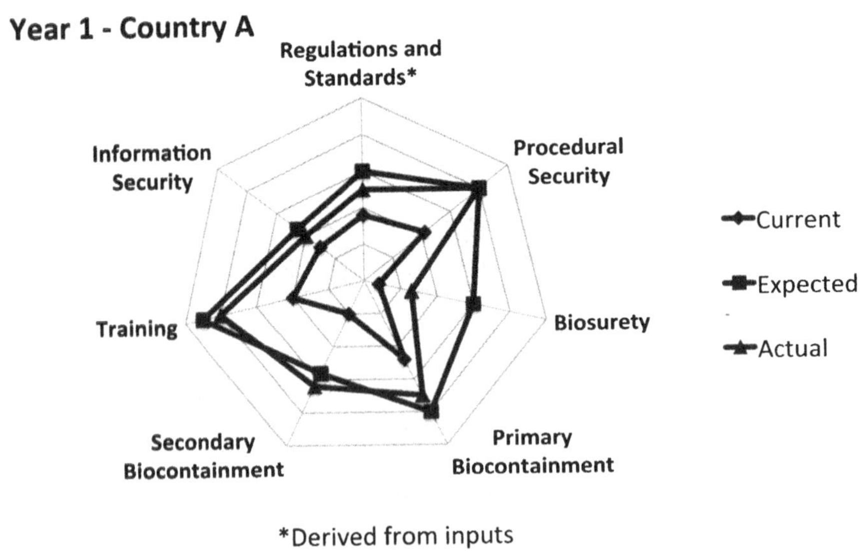

TABLE 14.7 Summary of impact, cost, and ratio of impact to cost for individual and vertical sets of laboratory biosafety activities

Domain		Activity 1	Activity 2	Activity 3	Projected	Total
Procedural Security	Impact	40	15	15	30	70
	Cost*	100	54	28	82	182
Biosurety	Impact	10	70	10	10	90
	Cost*	8	68	5	5	81
Physical Security	Impact	50	10	40	100	100
	Cost*	16	81	5	102	102
Information Security	Impact	15	40	35	75	90
	Cost*	35	4	84	88	123
Training	Impact	20	15	40	55	75
	Cost*	32	10	21	31	63

*Thousands $US

Calculation of impact versus cost ratios						
Procedural Security	Impact/Cost	0.40	0.28	0.54	0.37	0.38
Biosurety	Impact/Cost	**1.25**	**1.03**	**2.00**	**2.00**	**1.11**
Physical Security	Impact/Cost	**3.13**	0.12	**8.00**	0.98	0.98
Information Security	Impact/Cost	0.43	**10.00**	0.42	0.85	0.73
Training	Impact/Cost	0.15	**1.21**	0.10	0.10	0.30

chronologically within each defined set, a projected figure including the impact, cost, and impact/cost values is provided only for those activities that require completion. Note that cost estimates will be specific to the laboratory under assessment, as larger laboratories may require more resources to conduct the same activity.

Finally, an overview of current and expected progress can be displayed for each biosecurity domain, presuming each set of activities in each domain is conducted. This overview can later be expanded to include actual progress once an activity has been completed or is in progress (Figure 14.11).

LIMITATIONS

There are currently no models for monitoring and evaluating LBPs. Our work is new and, we hope, useful. Nevertheless, because there is no baseline, we acknowledge that this methodology should be pilot-tested and modified to fit the reality on the ground. Any analysis on the front and back end will likely have limitations, especially in sample size and incomplete data. Additional limitations might also include:

- Logic models conceived might not fit all types of laboratories (e.g., government vis-à-vis private).
- Indicator database includes those collected or developed for this assessment; there may be redundancies. Further, we may not have all indicators necessary to monitor certain aspects of the LBP.
- The cost to conduct an activity will vary from laboratory to laboratory based on the size of the lab and volume of work, among other factors. Determination of cost versus impact is therefore subject to the accuracy of the cost estimates factored into the analysis.
- PCA is a large-sample procedure. Insufficient sample size may compromise the statistical rigor of the analysis and fail to produce the desired result. Samples may be collected over time to achieve the necessary numbers for robust results.
- PCA does not perform well on binary data. If all or a substantial portion of the data within a given dataset is binary, other forms of analysis may be more appropriate.
- Preliminary assessments prior to meeting the required minimum will be more qualitative than quantitative, as model inputs will likely be based on subjective scoring of indicators.

CONCLUSION

A standardized monitoring and assessment protocol is essential for a laboratory director to ensure and maintain proper laboratory biosafety measures to protect laboratory staff and the public. As WHO member states work to meet IHR laboratory core capacity requirements, the conceptual framework, analytic processes, and data representation described in this chapter can be employed to determine how best to build those capacities based on qualitative and quantitative measures.

This conceptual framework systematically and rigorously defines the scope of laboratory biosafety and its component domains within five LBP sublogic models. Within each sublogic model, expected outcomes are mapped to stepwise activities. The indicator database provides measureable milestones to track the status or progress of an input, activity, output, or outcome. These indicators then map to various BSLs such that the assessment can be tailored to the BSL of a laboratory. Over time, the conceptual framework analysis can be refined to focus on the principal components of assessment, using PCA to simultaneously strengthen and simplify the tool and the monitoring and assessment process. Paired with accurate cost estimates, the collected indicator data should provide sound guidance and justification for conducting activities to increase laboratory biosafety capacities. This evaluation should allow laboratory directors to definitively set and execute activities to reach short-, intermediate-, and long-term goals according to their current status and appraisal of priorities given finite resources.

We recognize that these new concepts and analytic approaches require peer review, vetting, and pilot testing in a variety of developed and underserved global settings. And while we have started these processes, still more work is required. Evaluating the progress and challenges related to LBPs requires validation from the field. Both logic models and PCA are management tools that both describe work processes and define appropriate variables used for evaluation. However, both require pilot testing with real-world data. Therefore, we recommend a thorough pilot testing and validation of this new model. And while we have crafted a yardstick for measuring current LBP performance, as it were, we recommend that a panel of global laboratory experts be convened to discuss and adopt universal LBP standards.

References

1. Concept Plan for the Implementation of the National Biosurveillance Strategy for Human Health. Centers for Disease Control and Prevention (2010).
2. World Health Organization. Biorisk Management Laboratory Biosafety Guidance. WHO/CDS/EPR/2006.6 (2006).
3. Biosafety Levels. Hazard Awareness and Management Manual (HAMM): EH&S.
4. World Health Organization. Laboratory biosafety manual. 3rd ed. Geneva (CH)2004.
5. Bakanidze, L., Imnadze, P., Perkins, D. Biosafety and biosecurity as essential pillars of international health security and cross-cutting elements of biological nonproliferation. BMC Public Health. 10 Suppl 1:S12 (2010).
6. Organization WH. International Health Regulations. IHR Monitoring Framework. Checklist and Indicators for Monitoring Progress in the Development of IHR Core Capacities in States Parties. Geneva (2010).
7. Carr, K., Henchal, E.A., Wilhelmsen, C., Carr, B. Implementation of biosurety systems in a Department of Defense medical research laboratory. Biosecurity and bioterrorism: biodefense strategy, practice, and science. 2(1):7–16. Epub 2004/04/08 (2004).

8. Garaigordobil, M. Evaluation of a program to prevent political violence in the Basque conflict: effects on the capacity of empathy, anger management and the definition of peace. Gac Sanit. Epub 2012/01/31 (2012).
9. Jahrling, P., Rodak, C., Bray, M., Davey, R.T. Triage and management of accidental laboratory exposures to biosafety level-3 and -4 agents. Biosecur Bioterror. 7(2):135–43. Epub 2009/07/29 (2009).
10. Le Duc, J.W., Anderson, K., Bloom, M.E., Estep, J.E., Feldmann, H., Geisbert, J.B., et al. Framework for leadership and training of Biosafety Level 4 laboratory workers. Emerging Infectious Diseases.14(11):1685 (2008).
11. Lewis, M. Development CfG. Governance and corruption in public health care systems. Center for Global Development (2006).
12. Losinger, W.C., Bush, E.J., Hill, G.W., Smith, M.A., Garber, L.P., Rodriguez, J.M., et al. Design and implementation of the United States National Animal Health Monitoring System. National Swine Study. Preventive veterinary medicine. 34(2-3):147-59 (1998).
13. Miller, S.R., Bergmann, D. Biocontainment design considerations for biopharmaceutical facilities. Journal of Industrial Microbiology and Biotechnology. 199311(4): 223–34 (993).
14. Murray, C.J.L., Evans, D.B. Health systems performance assessment. World Health Organization (2003).
15. Radelet, S., Caines, K. The Global Fund to Fight AIDS, TB, and Malaria: Performance and Vision. A Report Prepared for the UK Department for International Development. London. (2005).
16. Rychetnik, L., Frommer, M., Hawe, P., Shiell, A. Criteria for evaluating evidence on public health interventions. Journal of Epidemiology and Community Health. 56(2):119 (2002).
17. Taboy, C., Chapman, W., Albetkova, A., Kennedy, S., Rayfield, M. Integrated Disease Investigations and Surveillance Planning: a systems approach to strengthening national surveillance and detection of events of public health importance in support of the International Health Regulations BMC Public Health. 10 (Suppl 1): S6 (2010). Accessed at: http://www.biomedcentral.com/1471-2458/10/S1/S6.
18. Wurapa, F., Afari, E., Ohuabunwo, C., Sackey, S., Clerk, C., Kwadje, S., et al. One Health concept for strengthening public health surveillance and response through Field Epidemiology and Laboratory Training in Ghana. Pan Afr Med J. 10 Supp 1:6. Epub 2012/03/01 (2011).
19. Zhao, J., Lama, M., Sarkar, S., Atun, R. Indicators Measuring the Performance of Malaria Programs Supported by the Global Fund in Asia" Progress and the Way Forward. PLoS ONE. 6(12):e28932 (2011).
20. Anderson, L., Petticrew, M., Rehfuess, E., Armstrong, R., Ueffing, E., Baker, P., et al. Using logic models to capture complexity in systematic reviews. Research Synthesis Methods. 2(1):33-42 (2011).
21. Bickman, L. The functions of program theory. In Bickman (ed.). Using program theory in evaluation: new directions for program evaluation. Jossey-Bass. San Francisco (1987).
22. Kaplan, S., and Garrett, K. The use of logic models by community-based initiatives. Evaluation and Program Planning. 28:167–72 (2005).
23. Agriculture Organization of the United Nations. Biosecurity Priority Area for Interdisciplinary Action FAO biosecurity toolkit. Food and Agriculture Organization (2007).

24. Report of the Trans-Federal Task Force on Optimizing Biosafety and Biocontainment Oversight (2009).
25. 73 Cp. The Possession, Use, and Transfer of Select Agents and Toxins (2005).
26. American Institute of Chemical Engineers. Center for Chemical Process Safety. Guidelines for process safety in bioprocess manufacturing facilities. Wiley-AIChE. Hoboken, N.J. (2011).
27. WHO Biorisk Management: Laboratory biosecurity guidance (2006).
28. CfDC. Comprehensive plan for epidemiologic surveillance. US Department of Health and Human Services CDC. Atlanta (1986).
29. Chosewood, L.C., Wilson, D.E. Centers for Disease Control and Prevention. National Institutes of Health. Biosafety in microbiological and biomedical laboratories, 5th ed. Deptartment of Health and Human Services. Public Health Service. Centers for Disease Control and Prevention. National Institutes of Health. Washington, DC (2009).
30. Wilson, D., Chosewood, L. Biosafety in microbiological and biomedical laboratories. Department of Health and Human Services. CDC/NIH, 5th ed. Washington, DC (2007).
31. Organisation for Economics Co-operation and Development. OECD best practice guidelines for biological resource centres. Paris (2007).
32. Health Canada. Laboratory biosafety guidelines, 3rd ed (lab manual). Ottawa (2004).
33. Organisation for Economics Co-operation and Development. OECD best practice guidelines for biological resource centres. Paris (2007).
34. University of Florida. Division of Environmental Health and Safety. Biological safety manual (lab manual). Gainesville (2008).
35. Yale University. Office of Environmental Health and Safety. Biological safety manual (lab manual). New Haven (2008).
36. Princeton University. Biological safety manual (lab manual). Princeton (2011).
37. Harvard University. Department of Chemistry and Chemical Biology. Laboratory safety manual (lab manual). Cambridge (2012).
38. Greenacre, M. Correspondence analysis of the Spanish National Health Survey. Gac Sanit.16(2):160–70 (2002).
39. Joliffe, I.T., and Morgan, B.J. Principal component analysis and exploratory factor analysis. Stat Methods Med Res.1(1):69–95 (1992).
40. Ott, L., and Longnecker, M. An introduction to statistical methods and data analysis, 5th ed. Duxbury. Pacific Grove, CA (2001).
41. Hu, F.B. Dietary pattern analysis: a new direction in nutritional epidemiology. Curr Opin Lipidol.13(1):3–9 (2002).
42. Abdi, H., and Williams, L. Principal Component Analysis. WIREs Comp Stat. 2(4): 433–59 (2010).
43. Jombart, T., Devillard, S., Balloux, F. Discriminant analysis of principal components: a new method for the analysis of genetically structured populations. BMC Genet. 11:94 (2010).
44. Ficetola, G., Marziali, L., Rossaro, B., De Bernardi, F., Padoa-Schioppa, E. Landscape-stream interactions and habitat conservations for amphibians. Ecol Appl. 21(4): 1272–82 (2011).
45. Panagiotakos, D.B., Pitsavos, C., Skoumas, Y., Stefanadis, C. The association between food patterns and the metabolic syndrome using principal components analysis: The ATTICA Study. J Am Diet Assoc. 107(6):979–87, quiz 97 (2007).

46. Porter, A., NC, N., Jin, X., Johnson, D., Roessner, J. High tech indicators: technology-based competitiveness of 33 nations. Georgia Institute of Technology, Technology. Policy and Assessment Center. Atlanta (2008).
47. Vyas, S., Kumaranayake, L. Constructing socio-economic status indices: how to use principal components analysis. Health Policy Plan.21(6):459–68 (2006).
48. Hatcher, L. A Step-by-Step Approach to Using the SAS 1 System. SAS Institute. Cary, NC (1994).

CHAPTER 15

The Future of Biosecurity: A Global Context

Ryan N. Burnette, J. Craig Reed, and Patricia Delarosa

INTRODUCTION

Biosecurity is both a concept, and a set of practices and principles composed of a variety of elements. The objectives of this book are to provide clarity on the facets of biosecurity, its applications to modern-day society, and its future biosecurity's global implications. Most importantly, this book frames the concepts linking biosecurity across global health, defense, research, agriculture, and medicine.

BIOSECURITY SUMMARIZED

Chapter 1 provided detailed definitions of biosecurity and its related terms. From its strict interpretation, biosecurity is defined as "concepts and practices that are used to secure the access of sensitive biological materials from persons or entities that should not have access."[1] This definition is appropriate to describe biosecurity in the context of a research laboratory but fails to encompass broader models of biosecurity used in agriculture, transportation, or public health (Chapters 8, 9, and 10). At a policy level, biosecurity is also defined as "a strategic and integrated approach that encompasses the policy and regulatory frameworks (including instruments and activities) that analyze and manage risks in the sectors of food safety, animal life and health, and plant life and health, including associated environmental risk."[2] In short, defining biosecurity to encapsulate single tasks, as well as larger concepts such as biodefense and bioterrorism, is no trivial task.

Biosafety and Biosecurity are Intertwined

There is a tangible correlation between biosafety and biosecurity, as is explored in Chapter 2. Biosafety has been codified into a set of internationally accepted practices and principles through the publication of multiple guideline documents by various governments and international organizations[3,4,5] and, through Codes of Federal Regulation (U.S. CFR). Regulations surrounding biosafety are still lacking, despite multiple incidents prompting researchers and legislators to take further action. In essence, biosafety serves as a set of practices and procedures to protect people and the environment from hazardous microorganisms that may negatively affect human, animal, or plant health. Biosecurity, on the other hand, is focused more on protecting the biological agents from individuals with malicious intent. At the level of laboratory practices, biosafety and biosecurity are intertwined and must be considered together.

Biosecurity Has Many Facets

Several chapters in this book provide a detailed analysis of the facets of biosecurity. We explored the requirements of physical infrastructure (Chapter 4) necessary for facilities to achieve rudimentary and advanced biosecurity measures. However, physical biosecurity alone is insufficient to address many aspects that jeopardize institutions housing valuable biological materials (VBM), such as "insider threats." Rather, physical biosecurity must be coupled with a sustainable and compliant operational biosecurity program.

Biosecurity, like biosafety, relies heavily on the use of a risk assessment. The fundamental concept of a risk assessment is a complete understanding of the VBM being manipulated and housed. Different VBMs present different risks to the end user. A primary example is the difference between a risk group 2 (RG2) organism (e,g,, hepatitis) and a risk group 3 (RG3) organism (e.g., hantavirus). While not exclusive, RG2 organisms are manipulated under BSL-2 conditions, and RG3 organisms are handled at BSL-3 organisms (Chapter 1). The primary consideration is that organisms manipulated in BSL-3 facilities are aerosol-transmissible—the determination of which BSL laboratory is required stems from the risk assessment. Further, an organism that is aerosol-transmissible may be more attractive to the would-be bioterrorist, as it is potentially more dangerous to the public—this determination is part of the threat assessment (Chapter 6). This interrelationship between the risk assessment and the threat assessment is a critical component of biosecurity. This book also provides many of the fundamentals as they apply to operational biosecurity (Chapter 5).

Operational biosecurity also employs control measures necessary to protect VBM from insider and outsider threats. The major components are the risk assessment, the threat assessment, administrative controls, and development of institutional biosecurity plans and programs. Personnel suitability and reliability (often referred to as "biosurety") programs, where individuals with access to dangerous or VBM are "screened" to determine if the individual is competent to have access to such materials, are also beginning to play a larger role in institutional biosecurity programs. Further, management of materials (inventory management) and information technology/security remain critical to effective biosecurity plans.

Biosecurity is not one component, or even one procedure. Rather, it is the culmination of understanding risks associated with dangerous materials or VBM, the threats presented to the VBM, as well as the physical and operational requirements to maintain a comprehensive biosecurity program. While biosecurity "recipes" will vary from institution to institution, the fundamentals described in this book can apply to any environment manipulating or housing dangerous materials or VBM.

Biosecurity Impacts Many Industries

It is obvious that major stockpiles of dangerous materials or VBM exist in research laboratories in the academic (universities), government (federal research labs) and private (biotechnology/ pharmaceutical) sectors. It is difficult to say whether biosecurity has received the necessary attention even in these obvious industries. However, much less attention has been given to biosecurity in the context of other industries, such as agriculture, transportation, and public/global health. This book provides a detailed review of biosecurity as it pertains to these tangential industries.

In particular, agriculture faces serious biosecurity threats (Chapter 8). It has major economic value, a concentrated infrastructure, and many points of potential weakness. For these reasons and others, agriculture remains at risk from both natural and malicious threats. Further, the agriculture industry has only recently attracted the attention of legislators.

Biosecurity Policy is Fragmented and Incomplete

It is unfair to say that biosecurity was not considered in the U.S. or other developed nations prior to 2001 (9/11, anthrax letters). However, awareness of biosecurity issues became more widespread in subsequent years. Since those attacks, new legislation fortifies research institutions and prevents a biological attack, as well as increasing general national security. This is probably best demonstrated by the creation of a new cabinet office, the Department of Homeland Security (Chapter 9).

Chapter 11 provided an in-depth review of current and future policy relating to biosecurity in the U.S. Despite the perceived breadth and depth of existing biosecurity policy, legislation remains fragmented and incomplete. Like biosafety, no standard exists for laboratory biosecurity. However, in 1999 the 4th edition of the CDC/NIH publication Biosafety in Microbiological and Biomedical Laboratories (BMBL) included guidance on biosecurity (Chapter 2). Today, the BMBL remains a guidance document, not a standard.

Laboratory stakeholders have experienced a greater awareness of biosecurity issues. As discussed previously, the interconnectedness of biosafety and biosecurity refers to "biorisk management" (Chapter 1). Certain guidance documents emphasizing laboratory biorisk management have likely made the concept of biosecurity palatable for laboratories in the U.S. and around the world. This is best exemplified by the European Committee for Standardization (CEN) Laboratory Biorisk Management 15793.[6] Crafted by stakeholders worldwide, this document represents guidance for biosafety and biosecurity to laboratories globally. To date, this document has been through pilot implementation projects in the U.S. and elsewhere.

Taken together, advances in legislation and policy have occurred to further biosecurity initiatives. Yet, few standards for institutions with biosecurity considerations exist. It is likely that unpredicted events, such as continued disease outbreaks or even malicious biological attacks, will be necessary to catalyze additional legislation.

A ONE WORLD/ONE HEALTH MODEL FOR BIOSECURITY

The global context of biosecurity is best presented as a component of the one world/one health model. The National Strategy for Countering Biological Threats calls for implementation of protection and prevention measures to counter the risks associated with the release of a biological agent.[7] However, in the WMD Center's Bio-Response Report Card the topic of prevention is not covered. This reflects a general consensus in the biodefense community that prevention strategies for biological threats are of limited value.[8] Prevention of bioterror events and infectious disease outbreaks are interrelated, and recent efforts in international biodefense strategy have included public health surveillance. Surveillance programs, however, are only one component of a prevention program and do not prioritize risk or effectively communicate risk mitigation strategies. In agriculture, the Food and Agriculture Organization (FAO) has identified three objectives for an animal health biosecurity program: "bio-exclusion" (elimination of disease), "biocontainment" (prevention of disease transmission), and strong cross-disciplinary communication. These objectives are accomplished using FAO biosecurity risk analysis as a tool to prioritize risk and to identify and communicate best practices. As discussed in Chapter 10, in public health the scope of a preventive biosafety and biosecurity program must encompass all disciplines in the life sciences, including biotechnology, biosafety, and biosecurity.

Prevention strategies in public health historically have had the highest value, lowest cost methods to prevent the spread of infectious disease.[9] Moreover, the lack of prevention strategies is linked to increased incidences of disease.[10] In the United States, the rapid control of the West Nile virus (discussed in Chapter 10) highlights the critical importance of rapid detection and effective communication in an outbreak response and alludes to our ability to respond to a biological attack (Chapter 10). Subsequent analysis of the outbreak, however, also suggests that, by the time an attack is realized, the pathogen will have already integrated or "established" into the environment.[11]

Why have prevention strategies been ignored or thought to be of little value in the effort to contain bioterrorism? The answer lies, in part, in the history of the development of the disciplines of biosecurity. The traditional focus in physical security has always been to protect a defined source. In the case of biosecurity, that source was identified as a repository, laboratory, or, in the case of an outbreak of animal disease, farm. Laws were rapidly enacted to secure dangerous pathogens in these locations.[12,13] Subsequently, a great effort was made to develop oversight systems to ensure that the highest consequence pathogens were secure. This was a reasonable first response, but the organisms are abundant in nature, and there is no mechanism to secure pathogens in the wild. Furthermore, a review of biological terrorist attacks reveals that bioterrorism incidents include but are not limited to the use of biological select agents and toxins (BSAT) in research laboratories. In fact, biological attacks have been linked to healthcare providers, and infectious disease outbreaks have been associated with training laboratories and agricultural and food processing facilities.[14,15,16,17] An analysis of biosafety and biosecurity guidelines also reveals the gaps in coverage for public and environmental health (Chapter 10). And the rise of the blood-borne pathogen epidemics in the 1970s to 1990s suggests that a focus on microbiology alone is not sufficient, and efforts must be extended to include all aspects of public health and life science research. These findings point to numerous weaknesses in our biosecurity network system that could be remedied by preventative biosafety and biosecurity measures.

The authors of the Report Card correctly identified the problems setting biosecurity apart from other security programs: "Deadly pathogens are readily available in nature, the knowledge and equipment required to weaponize these pathogens, and the capability needed to effectively deliver this weaponized material is dual-use standard procedure in the biotech/pharmaceutical/agriculture industries." This statement highlights the complexity of preventive public health biosafety and biosecurity program development and suggests such efforts may be too complex. Equally possible, however, is that the model we currently use is not applicable to the protection of a biological system. A better model is suggested by the One World/One Health paradigm, where an integrated approach to public health connects environmental, agricultural, microbiological, and clinical efforts.

Expansion of biosafety and biosecurity efforts beyond the laboratory is not a novel idea. The biosecurity community has leveraged public- and veterinary-disease surveillance efforts to define baseline infection levels and help in biological forensic investigations. These measures begin to address the role of prevention in public health. But while the role of epidemiology in biosecurity has been embraced, the roles of biosafety and biosecurity in the One Health initiative have been underestimated. Biosafety and biosecurity are disciplines, like epidemiology, and programs that run across disciplines in the One World/One Health model represent a common thread connecting these disparate disciplines. Only epidemiology, biosafety, and biosecurity have a communication component, and only biosafety and biosecurity include biotechnology, teaching, and security (Chapter 10). The most obvious advantage in the application of biosafety and biosecurity across the One World/One Health model would be the formation of a common language for communication of best practices and prevention strategies. Unorganized attempts have been made to establish definitions for biosafety and biosecurity within the disciplines of microbiology, public health, veterinary sciences, and ecology by the Centers for Disease Control, the World Health Organization, the OIE, and the Convention on Biological Diversity (Chapter 10). These efforts leave us with very different definitions of biosafety and biosecurity, resulting in different methods of risk analysis and communication of risk mitigation practices.[1, 18,19,20] Application of the One Health model to the Report Card would fundamentally redefine the field of communication and add a field for preventive biosafety and biosecurity research.

The One World/One Health model for biosecurity would require a cross-disciplinary, international research effort to establish methods and competencies in biosafety and biosecurity within each discipline. A decentralized, academic approach to biosecurity, as suggested in Chapter 10, provides a relatively low-cost model that is easily adapted to international systems and capabilities. In this model, individuals knowledgeable in discipline-specific biosecurity issues would work collaboratively with biosecurity experts to profile and prioritize emerging risks. Risk mitigation methods that work across societies in low-resource countries would need to be developed. Communication networks would need to be established. The establishment of a public health biosecurity system represents a large collaborative effort on the part of public health professionals, security experts, and academicians. The enormity of the task is far outweighed by the potential of the model to fill missing security functions in biosecurity. The simple act of redefining the model for biosecurity along the One Health model involves a wider spectrum of leaders in biosecurity and initiates a focus on whole-system risks—defining risks in discipline-specific terms. The effort should identify discipline-targeted proficiencies and accreditations that expand the biosafety and biosecurity effort and place an emphasis on prevention of terrorism.

In the Report Card, three strategic priorities are identified: (1) clear priorities and commitment and unity of effort, (2) mobilization of "whole of nation" response planning, and (3) sustained investment in purpose-driven science. In this context, the One World/One Health biosecurity model defines the scope of the "unity of effort," allows for selection of leaders within all the life science disciplines, and proves oversight of risks, vulnerabilities, and threats not covered by current legislation. It mobilizes all individuals in the life sciences from doctors and veterinarians to students in teaching laboratories. It expands roles and responsibilities for biosecurity to include all academic institutions with life science and allied health curricula, as well as students in these programs. Finally, it promotes sustained investment in purpose-driven science, defines biosafety and biosecurity as disciplines within the biological sciences, suggests novel science-based methodologies to elaborate system-wide vulnerabilities, and promotes the identification of targeted, high-impact, low-cost strategies through applied biosafety and biosecurity research.

FUTURE OF BIOSECURITY

The revised 5th edition of the BMBL provides a brief narrative on biosecurity. However, terms such as "threat assessment" and "suitability assessment" are never used, and the critical role of human resource personnel and appropriately trained security personnel in the screening of employment candidates is not discussed. In the past three years there has been a significant expansion and penetration of awareness as to how to effectively implement biosecurity. Now a federal guidance document (much like the BMBL) for biosecurity is needed for the industry. It should be a stand-alone edited volume that incorporates the growing body of biosecurity guidance. It should reemphasize the core principle that biosafety and biosecurity are inextricably intertwined; effective biosafety is essential for effective biosecurity and vice versa. Just as the current BMBL is written for consumption by a variety of professionals who work in containment laboratories and are involved in their daily operation and management, the focus of a biosecurity guidance document should be targeted to the same audience. The creation of this document should be led and supported by the CDC and NIH. Basic security terminology should be defined and the fundamental principles of biosecurity should be discussed, including:

- The significant difference between a biosafety risk assessment and a biosecurity risk assessment
- The role of threat assessment in the context of a biosecurity risk assessment
- The credentials and roles for each of the professionals involved in successful biosecurity implementation
- The framework for a successful biosecurity program
- The targeted violence process
- The benefits and drawbacks of suitability programs for employee prescreening and threat assessment investigations for employees who display threatening behavior

The challenges of biosecurity are not strictly limited to the U.S. Other countries struggle with biosecurity challenges. These challenges range from a lack of awareness to inadequate implementation. Many organizations supporting international scientific engagement efforts discovered that different countries not only place differing levels of emphasis on biological safety (to include little or none) but approaches that work well in one country may not work well in another.[21] The same is also true of biological security. As pointed out in 2011 by Heckert et al., the solution set falls into nine basic categories:[22]

1. The development of regulatory frameworks and guidelines or standards for biosecurity
2. Biosecurity awareness
3. Appropriate infrastructure
4. Equipment and services that support biosecurity
5. Management processes and administrative controls
6. Biosecurity curricula for security professionals
7. Biosecurity training for laboratory staff
8. Biosecurity associations providing professional competency and credentialing
9. Individual mentoring and organizational twinning.

All nine of these elements are necessary to successfully address global biological security challenges.

U.S. international scientific engagement programs also revealed that many societies had neither a position title nor a role in the organizational structure for a biosafety officer. This stemmed from a systemic perceived lack of value in having a single individual serve as the organization's coordinator or focal point for laboratory biosafety. As biosafety and biosecurity have experienced significant development and expansion in recent years, perhaps it is time to reflect on lessons our society has learned on distant shores and apply them to our situation. Perhaps it's time to consider the identification of a new security function in the organizational structures of our laboratories, a function, supported by the appropriate credentials and experience, with the title of "biosecurity officer." Just as the competencies for biosafety practitioners have been defined through an international consensus process managed by the European Committee of Standardization (CEN) and through a joint effort of the CDC and the Association of Public Health Laboratories, perhaps a similar set of recommended competencies for biosecurity should be developed.[23,24]

CONCLUSION

This book has examined biosecurity from its components to its impacts, not only in the research world but in other industries that have a stake in biosecurity. It is clear that biosecurity exists as a culmination of ideas, principles, practices, physical infrastructures, and legislative action. In short, biosecurity is complex.

History has demonstrated a reactionary trend in society against implementing additional measures to protect dangerous materials and VBM from potential threats. Granted, the complexity of the protection systems required provides for a difficult implementation process. Regardless, biosecurity remains under-addressed in many industries, such as agriculture and public/global health. In other areas, such as nanotechnology, biosecurity may have no current impact. However, this should not keep us from being proactive and planning for the future. Other industries and technologies will emerge, requiring the tenets of biosecurity. Our world and population will continue to combat naturally occurring disease. Developing nations will continue, albeit incrementally, to fortify their laboratories and health systems, potentially creating additional stockpiles of dangerous clinical samples—a process likely to go unchecked in the defense and security world for some time to come. Potential bioterrorists will continue to plot attacks and possibly attempt to act on those plots. There is no shortage of challenges presented to stakeholders of research, agriculture, defense, and other industries.

Provided that stakeholders work progressively together to continually evaluate biosecurity ideas and technologies, we may find ourselves better prepared to stave off naturally occurring diseases as well as potential biological attacks.

References

1. Meyerson, L.A., and Reaser, J.K. A unified definition of biosecurity. Science 295: 44 (2002).
2. Food and Agriculture Organization of the United Nations. Biosecurity for Agriculture and Food Production. Accessed at: http://www.fao.org/biosecurity/.
3. U.S. Department of Health and Human Services, Public Health Service, Centers for Disease Control and Prevention, National Institutes of Health. Biosafety in Microbiological and Biomedical Laboratories, 5th ed. In Chosewood, L, Wilson, D (eds). Washington, DC (2009).
4. World Health Organization. Laboratory Biosafety Manual, 3rd ed. Geneva (2004).
5. Health Canada. The Laboratory Biosafety Guidelines, 3rd ed. Ottowa (2004).
6. European Committee for Standardization. CEN Workshop Agreement: Laboratory Biorisk Management. CWA 15793 (2008). Accessed at: ftp://ftp.cenorm.be/CEN/Sectors/TCandWorkshops/Workshops/CWA15793_September2011.pdf.
7. Miller, J.E. National Strategy for Biological Threats. Encyclopedia of Bioterrorism Defense (2011).
8. WMD Center. The Bipartisan WMD Terrorism Research Center's bio-response report card: 21st century biological threats. Washington, D.C. (October 2011).
9. Bruce D. Prevention of Disease. Science. 60:109–124 (1924).
10. Hughes, J.M., Wilson, M.E., Pike, B.L., et al. The origin and prevention of pandemics. Clinical Infectious Diseases. 50:1636 (2010).

11. Hayes, E.B., Komar, N., Nasci, R.S., et al. Epidemiology and transmission dynamics of West Nile virus disease. Emerg Infect Dis. 11:1167–1173 (2005).
12. Centers for Disease Control and Prevention, Department of Health and Human Services. Possession, use, and transfer of select agents and toxins. Final rule. Fed Regist. 200873:61363–61366 (2008).
13. U.S. Department of Agriculture. Agricultural bioterrorism protection act of 2002: possession, use, and transfer of biological agents and toxins; interim and final rule (7 CFR Part 331). Fed Regist.67:76908–76938 (2002).
14. Török, T.J., Tauxe, R.V., Wise, R.P., et al. A large community outbreak of salmonellosis caused by intentional contamination of restaurant salad bars. JAMA. 278:389–395 (1997).
15. Kolavic, S.A., Kimura, A., Simons, S.L., et al. An outbreak of Shigella dysenteriae type 2 among laboratory workers due to intentional food contamination. JAMA. 278: 396–398 (1997).
16. Beutin, L., and Martin, A. Outbreak of Shiga toxin-producing Escherichia coli (STEC) 0104:H4 infection in Germany causes a paradigm shift with regard to human pathogenicity of STEC strains. J Food Prot. 75:408–418 (2012).
17. Centers for Disease Control and Prevention. National Center for Emerging and Zoonotic Infectious Diseases. Division of Foodborne, Waterborne, and Environmental Diseases. Investigation Announcement: Multistate Outbreak of Human Salmonella Typhimurium Infections Associated with Exposure to Clinical and Teaching Microbiology. Accessed at: http://www.cdc.gov/salmonella/typhimurium-laboratory/042711/index.html.
18. World Health Organization. Biorisk management: Laboratory biosecurity guidance (2006). Accessed at: www.who.int/csr/resources/publications/biosafety/WHO_CDS_EPR_2006_6/en/.
19. World Health Organization. Food and Agriculture Organization of the United Nations. Food safety risk analysis. FAO Food and Nutrition Paper 87 (doi 978-92-5-1056-4-2). Rome (2006).
20. Secretariat of the Convention on Biological Diversity. Cartagena Protocol on Biosafety to the Convention on Biological Diversity: text and annexes. Montreal (2000). Accessed at: http:\\www.biodiv.org.
21. Reed, J.C. Bridging the Gap: The International Federation of Biosafety Associations. J. Am. Biol. Safety Assoc. 15(1), 3–5 (2011).
22. Heckert, R.H., Reed, J.C., Gmuender, F.K., Ellis, M., and Tonui, W. International Biosafety and Biosecurity Challenges: Suggestions for Developing Sustainable Capacity in Low-resource Countries. J. Am. Biol. Safety Assoc. 16(4), 223–230 (2011).
23. European Committee for Standardization Workshop Agreement 16393. Biosafety Professional Competence. European Committee for Standardization. Brussels (2012). Available at: ftp://ftp.cen.eu/CEN/Sectors/List/ICT/Workshops/CWA%2016393.pdf.
24. Delany, J.R., Pantella, M.A., Rodriguez, J.A., Shah, K.V., Baxley, K.P., and Holmes, D.E. Centers for Disease Control and Prevention and the Association of Public Health Laboratories. Guidelines for Biosafety Laboratory Competency. Morbidity and Mortality Weekly Report Supplement. Vol. 60 (April 15, 2011).

U.S. Food Defense since 9/11: Public Sector Initiatives and Programs

Henry S. Parker

REGULATORY AND POLICY FRAMEWORK

Between 2001 and early 2004, the U.S. federal government established and implemented a substantial regulatory and policy framework to address terrorism. The framework directly and indirectly strengthened the nation's ability to defend against agroterrorism. Key initiatives, legislation, and presidential pirectives are as follows:

- Office of Homeland Security. Less than a month after the 9/11 attacks, President George W. Bush issued an Executive Order that established an Office of Homeland Security.[1] The mission of the office was to "develop and coordinate the implementation of a comprehensive national strategy to secure the United States from terrorist threats or attack."
- National Strategy for Homeland Security. The strategy was signed by President Bush on July 16, 2002. Its purpose was to "mobilize and organize our nation to secure the U.S. homeland from terrorist attacks."[2] The strategy identified, for the first time, agriculture and food as critical infrastructure sectors and designated the United States Department of Agriculture (USDA) as the lead agency for protecting the agriculture sector. The strategy also called for the establishment of a cabinet-level Department of Homeland Security. The strategy emphasized that the private sector, as the owner of the vast majority of the nation's critical infrastructure, had the primary role and responsibility for protecting it.
- Homeland Security Act of 2002. This act, Public Law 107-296 of November 25, 2002, established the Department of Homeland Security (DHS) with the mission to prevent terrorist attacks in the U.S., reduce

vulnerability to attacks; minimize damage, assist in recovery from attacks, and coordinate federal, state, local, and private efforts in homeland security. The act included two provisions related to food and agriculture: it transferred certain agricultural import and entry inspection functions from USDA to DHS, and it transferred operating authority for USDA's Plum Island Animal Disease Center from USDA to DHS.[3]

- Public Health Security and Bio-Terrorism Preparedness and Response Act of 2002. Known as the "Bioterrorism Act," this legislation amended Public Health Service Act (42 U.S.C. 201) to strengthen the nation's ability to defends against bioterrorism. The act contained five titles, including titles pertinent to food and agriculture.[4]
- The Animal Health Protection Act of 2002, as amended through 2008 (7 U.S.C. 8310). The act established a consolidated statutory framework for all animal quarantine and related laws and provided statutory authority for the Animal and Plant Health Inspection Service (APHIS) of USDA to act decisively and swiftly to protect animal health.[5]
- The Food Safety Modernization Act of 2010 (P.L. 111-353). Signed into law on January 4, 2011, the Act gave FDA new authorities to protect the U.S. food supply by placing more emphasis on the prevention of contamination, rather than response to it.[6]
- Homeland Security Presidential Directives (HSPDs) and Presidential Decision Directives (PDDs). In addition to new national homeland security legislation, the Executive Branch has issued a number of policy directives and guiding principles to establish a framework and direction to improve the nation's ability to defend against terrorist attacks, major disasters, and other emergencies. Many of these are directly applicable to food defense.
    - HSPD-5. Management of Domestic Incidents Directive (February 28, 2003). This directive called for the establishment of a national incident management system (NIMS) and the development of a framework for a national response plan, with responsibility delegated to every federal agency head.[7] The national response plan was superseded by the national response framework on March 22, 2008 (see below).
    - HSPD-7. Critical Infrastructure Identification, Prioritization, and Protection Directive (December 17, 2003). This directive formally recognized, for the first time, that agriculture and food are critical national infrastructures. It designated sector-specific federal agencies and called for collaboration and coordination with the private sector in the development of plans and actions to protect the sectors.[8]

- HSPD-8. National Preparedness Directive (December 17, 2003). The directive called for establishment of a national "all-hazards" preparedness goal, provided federal assistance to state and local governments and first responders, established and maintained a comprehensive training program, and encouraged citizen participation.[9] The directive was replaced by PPD-8 in the Obama administration (see below).
- HSPD-9. Defense of United States Agriculture and Food Directive (January 30, 2004). The Directive affirmed that "It is the policy of the United States to protect the agriculture and food system from terrorist attacks, major disasters, and other emergencies ..." It established an interagency framework for the protection of agriculture and food and designated DHS as the overall coordinator. The directive served as the framework for a comprehensive, interagency approach to address awareness and warning, vulnerability assessments, mitigation strategies, response planning and recovery, outreach and professional development, and research and development.[10]
- HSPD-10. Biodefense for the 21st Century Directive (April 28, 2004). The directive established the following framework for federal departments and agencies to protect critical national interests (including food and agriculture) against the threat of bioterrorism: threat awareness, prevention and protection, surveillance and detection, and response and recovery.[11]
- PPD-8. National Preparedness Directive (March 30, 2011). The directive replaced HSPD-8. While similar to HSPD-8, PPD-8 defined the following terms: preparedness, security, resilience, prevention, protection, mitigation, response, and recovery. The directive gave specific timelines for the submission of the national preparedness goal (180 days), a description of a national preparedness system (240 days), and submission of an annual national preparedness report.[12]

- National Response Framework (January 2008). The national response framework (NRF) is a comprehensive guide to a national response to all hazards. The framework links government and non-government organizations and the private sector in a coordinated approach (operating through the national incident management system) to the management of serious incidents. The framework identified 15 emergency support functions (ESFs), with designated responsibilities and responsible federal agencies. These are compiled in a separate annex to the framework. ESF #11 addresses agriculture and natural resources, coordinated by the USDA.[13]

- National Infrastructure Protection Plan (2009). As called for in HSPD-7, and underscored by HSPD-8 and PPD-8, the national infrastructure protection plan (NIPP) lays out a comprehensive approach to integrated protection of the nation's critical infrastructure and key resources (CIKR) through a single, coordinated national effort. It identifies 18 separate CIKR sectors, including food and agriculture, incorporates sector-specific plans for each, establishes a risk management framework, and defines specific roles and responsibilities for each sector-specific agency.[14]

GOVERNMENT INITIATIVES AND ACTIONS

Since 9/11 federal, state, and local governments, often in partnership with the private sector, have undertaken a number of initiatives and actions to address the threat of agroterrorism and to protect U.S. food and agriculture from both unintentional and deliberate disruption. These efforts have used the NIPP risk management framework and the components of HSPD-9 (see above) as a template. The NIPP risk management framework incorporated the following steps with feedback loops between and among the steps:

- Set security goals
- Identify assets, systems, networks, and functions
- Assess risks
- Prioritize
- Implement protective programs
- Measure effectiveness

USDA's food defense activities and programs are primarily conducted by the Animal and Plant Health Inspection Service (APHIS; plant and animal health); the Agricultural Research Service (ARS; research); the National Institute of Food and Agriculture (NIFA; research and education); and the Food Safety and Inspection Service (FSIS; food safety). In the Department of Health and Human Services (DHHS), food defense is the primary responsibility of the Food and Drug Administration (FDA). Working closely with USDA, DHS has consolidated agriculture and food defense activities in the Food, Agriculture, and Veterinary Defense Branch of its Office of Health Affairs. The department's biodefense research and education efforts are housed in its Science and Technology Directorate.[15]

Principal accomplishments, as related to the specific components of HSPD-9, are as follows (note that many of these actions are cross-cutting among HSPD-9 components):

- Awareness and warning
  - FSIS and FDA jointly developed a food emergency response network (FERN) that "integrates the nation's food-testing laboratories at the local, state, and federal levels into a network that is able to respond to emergencies involving biological, chemical, or radiological contamination of food." It is currently established in all 50 states and Puerto Rico and involves 37 federal labs, 116 state and agricultural labs, and 17 city and county labs.[16] Building on FERN, FSIS, FDA, and DHS have partnered with the National Association of State Departments of Agriculture (NASDA) to develop a model food emergency response plan (NASDA-FERP) template for states.[15]
  - USDA has established a national animal health laboratory network (NAHLN) that involves a cooperative effort between two USDA agencies—APHIS and NIFA—and the American Association of Veterinary Laboratory Diagnosticians. NAHLN's mission is to "provide accessible, timely, accurate, and consistent animal disease laboratory services nationwide, provide laboratory data to meet epidemiological and disease reporting needs, maintain the capacity and capability to provide laboratory services in support of responses to foreign animal disease outbreaks or other adverse animal health events, and focus on diseases of livestock (exotic, zoonotic, and emerging diseases), while including diseases of nonlivestock species."[17]
  - With funding support from USDA, a national plant diagnostic network (NPDN) was established in 2002. The mission of the NPDN is to focus on "the early detection, accurate diagnosis, and rapid communications needed to help mitigate the impact of endemic, emerging, and exotic pathogens and pests that attack agricultural, forest, and landscape plants in the U.S." The mission is accomplished "through a coordinated network of diagnostic laboratories and experts at land grant universities, state departments of agriculture, and industry developing and deploying regionally and nationally coordinated programs in diagnostics, training and education, and response." The NPDN's key role in the early detection of Asian soybean rust in the southeastern U.S. in 2004 was discussed above.[18]
  - It is important to be able to trace individual livestock in order to track diseased and at-risk animals to ensure a rapid response in the event of a disease outbreak. After unsuccessful attempts to establish a national animal identification system, USDA issued a proposed rule, on August 9, 2011, to improve the traceability of livestock moving between states during an animal disease incident. The department has solicited comments on the proposed rule.[19]

  - DHS has included biological agents of concern to the food and agriculture community in its BioWatch 24/7/365 biodetection capability.[15]
  - The mission of the National Biosurveillance Integration Center (NBIC) in DHS is to "enhance the capability of the federal government to rapidly identify, characterize, localize, and track a biological event of national concern." NBIC integrates and analyzes data including that related to food and agriculture.[20]
- Vulnerability assessments
  - As called for in the NIPP, and under DHS leadership, federal agencies are working with state and local agencies to assist states in identifying and assessing critical infrastructures and key resources (CIKR) related to the food and agriculture sector. To accomplish this, the participants employ a food and agriculture criticality assessment tool (FASCAT). The assessment is an important step in the NIPP risk management framework, discussed above.[21]
  - A strategic partnership program on agroterrorism (SPPA) is underway. Building on information developed by FASCAT, USDA, FDA, DHS, and the Federal Bureau of Investigation (FBI), these groups are in collaboration with states and the private sector to better understand food and agriculture sector CIKR, identify early warning indicators of potential attacks, and identify R&D needs and other strategies to prepare for and mitigate attacks. Thirty-six SPPA assessments have been completed to date.[22]
  - USDA and FDA have jointly developed a computerized risk assessment tool, called CARVER + Shock, to identify areas in the food and agriculture sector that may be vulnerable to attack. Based on a tool originally developed by the U.S. military, Carver + Shock "evaluates potential vulnerabilities in the supply chain of different foods and food processes."[23] The acronym stands for different attributes that help evaluate potential targets for an attack, as follows (quoting directly from the FDA web site for Carver + Shock: Enhancing Food Defense).[23]

- "Criticality: what impact would an attack have on public health and the economy?
- Accessibility: how easily can a terrorist access a target?
- Recuperability: how well could a system recover from an attack?
- Vulnerability: how easily could an attack be accomplished?
- Effect: what would be the direct loss from an attack, as measured by loss in production?
- Recognizability: how easily could a terrorist identify a target?"

- Mitigation Strategies. HSPD-9 identifies two key objectives for mitigation strategies: "... protect vulnerable critical nodes of production or processing from the introduction of diseases, pests, or poisonous agents ..." and "expand development of common screening and inspection procedures for agriculture and food items entering the United States and...maximize effective domestic inspection activities for food items within the United States."[10] Measures to date have included guidance and training for the food and agriculture industry, including the SPPA initiative described above. In addition, the provisions of the Bioterrorism Act of 2002 and the Food Safety Modernization Act of 2010 have given FDA substantial new authority to prevent contamination of the U.S. food supply (see earlier discussion).
- Response planning and recovery. Response planning and recovery efforts related to food defense have responded to the Directives of HSPD-8 and PPD-8 through the incorporation of the food and agriculture sector into the national incident management system and the national infrastructure protection plan (though development of the food and agriculture sector specific plan) and through the development and implementation of ESF #11 of the national response framework. These efforts were described earlier in the chapter. Additional critically important actions have included:
  - Development of a national veterinary stockpile (NVS) of vaccines, antivirals, and therapeutic products to rapidly respond to an animal disease outbreak. Currently funded at $3.8 million annually, the NVS is still in early stages of development.[24]
  - Development and implementation of a national plant disease recovery system (NPDRS) to ensure that the nation can rapidly respond to a major outbreak of plant diseases of commercially important crops. The goals are to develop disease-resistant germplasm and pesticide control methods that can be rapidly deployed. To date, under the direction of USDA-ARS, plans have been completed for approximately 14 commercial species, and plans for a number of other species are under development.[25]
  - Federal guidance and support to the agricultural industry in the development and implementation of food defense plans. USDA-FSIS has developed comprehensive guidance and a template for food defense for individual companies. Components include: outside security measures, inside security measures, personnel security measures, incident response security measures, and a list of tools or possible security measures for food defense.[26] The agency has encouraged establishments to undergo an annual food defense plan review that can be certified by FSIS. USDA-FSIS has provided more targeted guidance

for specific food industry sectors. An example is guidance developed for the meat and poultry industries ("Developing a Food Defense Plan for Meat and Poultry Slaughter and Processing Plants").[27]

- Outreach and professional development.
  - A number of U.S. colleges and universities, including the Georgetown University Medical Center and George Mason University, now offer courses or programs in agroterrorism, agrosecurity, food defense, and related subjects.
  - The extension disaster education network (EDEN) is a nationwide, coordinated program based at U.S. land grant universities. Its mission is to help deliver information and educational resources to citizens concerned or impacted by disasters, including acts of terrorism. EDEN also provides online materials for courses such as plant biosecurity management, protecting America's food supply, and animal agrosecurity and emergency management.[28]
  - Federal agencies, including the Department of Defense, have conducted multi-agency, multistakeholder exercises related to food defense. Noteworthy among these is the crimson series, conducted between September 2002 and January 2003.[29] There were three separate exercises, developed by the ANSER Institute and sponsored by USDA: crimson sky (in cooperation with the North Carolina Department of Agriculture and DOD) and crimson guard focused on foreign animal disease outbreaks in the U.S., while crimson winter addressed a deliberate attack on the U.S. food supply. In June 2002 and February 2003, two silent prairie exercises, sponsored by the National Defense University, simulated an agroterrorism attack during a time when the military was undergoing mass deployment to overseas theaters.[29] Several states, including Georgia, North Carolina, Kansas, and Texas, have also conducted agroterrorism exercises with the National Guard as a player.[29] The scenario for several of these exercises was the deliberate introduction of foot and mouth disease.
  - Under the coordination of the National Center for Food Protection and Defense, federal, state, and local agencies have established FoodSHIELD, "a web-based system for communication, coordination, education, and training among the nation's food and agriculture sectors."[30]
- Research and development: Federal support for food defense R&D since 9/11 has been primarily dedicated to biodefense. Total U.S. government civilian biodefense funding since fiscal year (FY) 2001 has been $66.9 billion, allocated among HHS, DOD (civilian programs), DHS, EPA, the Department of Commerce, the Department of State, the National Science Foundation, and USDA.[31] The available funding data

do not break out amounts specifically focused on R&D. Further, many of the programs included under the biodefense funding categories also support non-biodefense objectives.

- The federal budget estimate for biodefense funding in FY 2011 was $5.5 billion, of which $3.8 billion was appropriated for HHS, $0.487 billion for DHS, and $90 million for USDA. The administration's biodefense funding request for 2012 was $6.4 billion, of which $4.48 billion was for HHS; $1.0 billion for DHS; and $106 million for USDA, none of which was specifically identified as supporting R&D.[31]
- The agriculture and food sector specific plan under the national infrastructure protection plan established research priorities for food defense, as follows:[32,33]
  - Animal health
    - Identify threats
    - Develop new and improved detection, diagnostic, and analytic methods and technologies
    - Develop vaccines and therapeutics (antivirals)
    - Control insect vectors of zoonotic diseases
    - Develop alternative treatment options for disease outbreaks
  - Plant/crop health
    - Identify existing and emerging plant pathogens for economically important species
    - Improve disease containment methodologies
    - Develop improved methods for disposing of potentially infectious materials
    - Identify and develop disease-resistant varieties (germplasm screening)
    - Determine host range and persistence of priority plant pathogens
  - Food safety
    - Develop and validate better identification and detection methods for potential threat agents (from farm to table)
    - Better understanding of the viability or survivability of threat agents in food matrices
    - Develop new and improved methods to inactivate pathogens and harmful chemicals
  - Food distribution
    - Develop improved technologies for infrastructure protection (detection and sensor systems, protection and prevention systems,

entry and access portals, advanced infrastructure architectures, human and social issues)

- Overarching
  - Improved disease surveillance
  - Improved knowledge of disease organisms and their reservoirs, transmission cycles, and methods to block transmission
  - Improved epidemiology

It was difficult to obtain precise information about federal agrodefense R&D funding. Despite this, it is clear that, since 9/11, the federal government has strengthened agency programs and federal partnerships with U.S. universities to advance research related to food defense. In USDA, research is conducted in-house by ARS and the economics research service (ERS). ARS research, conducted by federal scientists in government facilities, takes place in more than 90 research locations. The research is carried out by approximately 2200 scientists who work on 800 projects embedded in 18 national programs. The annual budget for ARS research is $1.1 billion.[34]

Research at several USDA laboratories specifically targets food defense in three broad areas: animal health, plant diseases and crop protection and quarantine, and food safety. Principal USDA laboratories for animal health research related to food defense are:

- The Plum Island Animal Disease Center, located on an island near Orient Point, NY. Administered by DHS, the lab conducts research on foreign animal diseases (FADs). ARS scientists in the foreign animal disease research unit study foot and mouth disease (this is the only location in the U.S. authorized to work with the FMD virus), classical swine fever; African swine fever, and other FADs as required. ARS scientists work closely with DHS scientists and with personnel at the APHIS foreign animal disease diagnostic laboratory (FADDL), co-located at the center. The FADDL is part of the APHIS national veterinary services laboratories (NVSL), whose mission is to "provid(e) diagnostic services, reagents, and training in world-class facilities; (respond) to animal health emergencies; take an active role in managing the national animal health laboratory network ...; (and) serve as an international reference laboratory."[19]
- The National Animal Disease Center (NADC), Ames, Iowa. The NADC mission is to "conduct basic and applied research on selected diseases of economic importance to the U.S. livestock and poultry industries."[35] Co-located at NADC, APHIS scientists from the NVSL operate a diagnostic bacteriology laboratory and a diagnostic virology laboratory.

- The ARS arthropod-borne animal disease research unit in Manhattan Kansas (U.S. Department of Agriculture, Animal and Plant Health Inspection Service, 2012d).
- The ARS southeast poultry research laboratory (SEPRL), Athens, Georgia.[36]
- USDA-FSIS operates three U.S. regulatory field services laboratories (Athens, Georgia; St. Louis, Missouri; and Alameda, California). The mission of the labs is to "coordinate and conduct laboratory analytical services in support of the agency's farm-to-table strategies in the disciplines of chemistry, microbiology, and pathology for food safety in meat, poultry, and egg products."[37]

In addition, APHIS carries out development, diagnostic, and demonstration programs in support of research in close cooperation with ARS. APHIS and ARS programs are co-located at the Plum Island Animal Disease Center and at USDA's National Animal Disease Center in Ames, Iowa.[38]

Extramural research in USDA is funded through the National Institute for Food and Agriculture and primarily supports land grant universities (LGUs). LGUs also receive research support from states through allocations of federal formula funds, as authorized by the Hatch Act of 1887.

DHS also provides substantial support for food defense, in close cooperation with USDA, through its science and technology (S&T) directorate. The mission of S&T's Chemical and Biological Division, Agricultural Defense Branch, is to "enhance current capabilities and develop state-of-the-art countermeasures for high-priority foreign animal diseases."[15] Focus areas include: vaccines and diagnostics, disease simulation and analysis tools, and agricultural screening tools. DHS supports agrodefense research in universities through a variety of grants and cooperative agreements and through recently established DHS centers of excellence. Two of these centers directly support food defense research: the National Center for Foreign Animal and Zoonotic Disease Defense at Texas A&M University[39] and the National Center for Food Protection and Defense at the University of Minnesota.[40]

Since 9/11, there has been substantial effort to improve and expand federal research facilities and biological containment laboratories in support of food defense. Notable among these initiatives is a FSIS biosafety level 3 (BSL-3) laboratory established in 2004 in Athens, Georgia, to carry out studies on a variety of biological threat agents of concern to food safety.[41] Of even greater potential impact, a new national bio- and agrodefense facility (NBAF) is under development in Manhattan, Kansas. The NBAF will replace the Plum Island Animal Disease Center. Its mission will be to "protect the nation's animal agriculture, food supply, and public health from natural or intentional outbreaks of foreign, emerging, and zoonotic (animal to human) diseases."[15]

In addition, with substantial congressional appropriations, the National Interagency Confederation for Biological Research (NICBR) is developing an interagency biodefense campus at Fort Detrick, Maryland. Major tenants include the Department of the Army (U.S. Army Medical Research Institute for Infectious Diseases), DHS (National Biodefense Analysis and Countermeasures Center),[42] the Centers for Disease Control and Prevention, the National Cancer Institute, the National Institute of Allergy and Infectious Diseases, and USDA (the ARS foreign disease-weed science research unit). Several of these laboratories in place or under development directly or indirectly include food defense in their research mission. The ARS laboratory operates a P-3 plant pathogen laboratory and greenhouse containment facilities to safely study infectious plant diseases.[43]

CONCLUSION

The private sector is the nation's first line of defense against deliberate or unintentional disruptions to the agriculture and food sector, and it bears the ultimate responsibility for protecting the food supply from farm to fork. The food and agriculture "private sector" consists of more than just private companies. It also includes "private voluntary organizations" (PVOs); private research, development, and education entities (including colleges and universities); commodity organizations and advocacy groups; professional organizations such as the American Veterinary Medical Association, the American Society of Animal Science, and the American Seed Trade Association; and other nongovernmental stakeholders.[32]

Given a large diversity of focal areas, agendas, and goals, the food and agriculture private sector has a wide variety of interacting interests, some of which are complementary and some of which are in conflict. These include economic viability (profits); customer satisfaction (mainly related to product quality, variety, appeal, and affordability); health and nutrition; animal welfare; environmental sustainability; use and allocation of finite resources (including land, water, and fossil fuels); food security; food safety; and food defense.[32] Reconciliation of these different interests as they relate to policy and programs in support of food defense can be a challenging proposition.

Still, the private sector, led by the food and agriculture industry, has made substantial progress to protect its products since 9/11. Much of this progress has resulted from close cooperation with government agencies, particularly through the application of CARVER + Shock for vulnerability assessments for specific companies; the strategic partnership program agroterrorism (SPPA; see description earlier in the chapter); and the food and agriculture sector coordinating council (FASCC), under the aegis of the national infrastructure protection plan. The FASCC is primarily represented by producers; processors, and manufacturers; restaurants and foodservice associations; retail associations; warehousing and logistics associations; and agriculture production inputs and services firms and associations.[22] This public-private partnership has helped the private sector to identify, assess, and manage risks; set security goals; identify critical assets, systems, networks, and functions; prioritize assets to be protected; implement protective programs; and measure the effectiveness of food defense programs. On April 17, 2012, the FASCC sponsored an interactive food defense workshop with government and industry representative at the 2012 food safety Summit. The workshop consisted of a comprehensive overview of food defense resources and tools of the federal government; it enabled industry leaders to share information on how they have incorporated food defense goals into their existing operations and provided an update on the status of key food defense elements of the Food Safety Modernization Act of 2011.[44]

The private sector has also invested heavily in research related to food defense. As stated in the food and agriculture sector specific plan, "Private industry hosts its own R&D activities where it sees a competitive advantage, where a gap must be addressed, or for other reasons. This work is typically conducted to benefit a particular company or industry and occurs independent of government activities."[33] The amount of private-sector funding for agriculture R&D is roughly equivalent to that of the public sector, though the information generated is often proprietary. In fact, between 1950 and 1990 private-sector funding for agriculture research increased faster that public-sector funding (growth rates have been equivalent since 1990).[45]

Since 9/11, a number of major companies in the food and agriculture sector have developed comprehensive biosecurity plans and programs that address both unintentional and deliberate disruption to their operations and that simultaneously cover both food safety and food defense. The approach taken by Perdue Farms provides an excellent example.[46] The company brought in an independent consulting team to conduct a biosecurity risk assessment to identify gaps and issues related to biosecurity in internal operations and in contract poultry growers. The team identified three major issues that required attention: (1) outsider access to buildings, products, and food/feed ingredients; (2) the potential for biological contamination; and (3) the firm's ability to respond to a major biosecurity or biosafety crisis. The team provided a number of specific recommendations to address these issues. Perdue subsequently implemented a comprehensive biosecurity program to respond to the recommendations. A key element was the development of a biosecurity risk management tool that allowed the company to quantitatively evaluate biosecurity risks at specific locations (both at company operating sites and at contract grower locations).

While private-sector entities can develop and implement many policies and programs in support of biosecurity in their operations, they can't do it all—government support is critical in a number of areas. These include:[32]

- Reconcile conflicting interests among different private-sector organizations
- Encourage partnership (including international) approaches (e.g., shared information; standardization of protocols)
- Develop a reasonable, less burdensome regulatory framework
- provide law enforcement
- Improve access to critical, relevant intelligence information
- Assist in response and consequence management (including disaster relief programs)
- Support education, training, and guidance
- Fund research, development, and demonstration related to food defense

References

1. The White House. Executive Order Establishing the Office of Homeland Security and the Homeland Security Council (October 8, 2001). Available online at: http://georgewbush-whitehouse.archives.gov/news/releases/2001/10/20011008-2.html.
2. The White House Office of Homeland Security. National Strategy for Homeland Security (July 2002). Available online at: http://www.ncs.gov/library/policy_docs/nat_strat_hls.pdf.
3. United States Congress. The Homeland Security Act of 2002. Public Law 107-296 (November 25, 200. Available online at: http://www.dhs.gov/xlibrary/assets/hr_5005_enr.pdf.
4. United States Congress. Public Health Security and Bio-Terrorism Preparedness and Response Act of 2002. Public Law 107-188 (June 12, 2002). Available online at: http://www.gpo.gov/fdsys/pkg/PLAW-107publ188/pdf/PLAW-107publ188.pdf.
5. United States Congress. Animal Health Protection Act. Public Law 107-171 (May 13, 2002), as amended through P.L. 110-246 (effective May 22, 2008).

6. United States Congress. Food Safety Modernization Act. Public Law 111-353 (January 4, 2011). Available online at: http://www.fda.gov/Food/FoodSafety/FSMA/ucm247548.htm.
7. The White House. Homeland Security Directive 5: Directive on Management of Domestic Incidents (February 28, 2003). Available online at: http://www.hsdl.org/?collection/stratpol&id=pd&pid=gwb.
8. The White House. Homeland Security Directive 7: Critical Infrastructure Identification, Prioritization, and Protection (December 17, 2003). Available online at: http://www.hsdl.org/?collection/stratpol&id=pd&pid=gwb.
9. The White House. Homeland Security Directive 8: National Preparedness (December 17, 2003). Available online at: http://www.hsdl.org/?collection/stratpol&id=pd&pid=gwb.
10. The White House. Homeland Security Directive 9: Defense of United States Agriculture and Food (January 30, 2004). Available online at: http://www.hsdl.org/?collection/stratpol&id=pd&pid=gwb.
11. The White House. Homeland Security Directive 10: Biodefense for the 21st Century (April 28, 2004). Available online at: http://www.hsdl.org/?collection/stratpol&id=pd&pid=gwb.
12. The White House. National Preparedness (March 30, 2011). Available online at: http://www.hsdl.org/?collection/stratpol&id=pd&pid=bo.
13. Federal Emergency Management Agency. National Response Framework (2008). Available online at: http://www.fema.gov/national-response-framework.
14. U.S. Department of Homeland Security. National Infrastructure Protection Plan (2009). Available online at: http://www.dhs.gov/national-infrastructure-protection-plan.
15. Meckes, D. Protecting the Agriculture and Food Sector—A Department of Homeland Security Priority (2009). Presentation to Georgetown University Medical Center course in agroterrorism (January 13, 2012).
16. Kelley, L. Food Defense: Protecting the Food Supply. Presentation to Georgetown University Medical Center course in Agroterrorism (February 10, 2012).
17. U.S. Department of Agriculture. Animal and Plant Health Inspection Service. National Animal Health Laboratory Network (2012). Available online at: http://www.aphis.usda.gov/animal_health/nahln/.
18. National Plant Diagnostic Network. Welcome to the National Plant Diagnostic Network Website (2012). Available online at: http://www.npdn.org/.
19. U.S. Department of Agriculture. Animal and Plant Health Inspection Service. Animal Disease Traceability (2012). Available online at: http://www.aphis.usda.gov/traceability/.
20. U.S. Department of Homeland Security. National Biosurveillance Integration Center (2012). Available online at: http://www.dhs.gov/national-biosurveillance-integration-center.
21. National Center for Food Protection and Defense. FASCAT: Food and Ag Sector Criticality Assessment Tool, Guidance Document (undated). Available online at: http://www.foodshield.org/fascat/docs/v3guidance.pdf.
22. Pulz, J. Countering the Threat: Federal Agency Perspectives. Presentation to Georgetown University Medical Center course in agroterrorism (January 27, 2012).
23. U.S. Food and Drug Administration. Carver + Shock: Enhancing Food Defense (2012). Available online at: http://www.fda.gov/ForConsumers/ConsumerUpdates/ucm094560.htm.

24. Gay, C. Foreign Animal Diseases Research and Biodefense. Presentation to Georgetown University Medical Center course in agroterrorism (March 22, 2012).
25. U.S. Department of Agriculture. Agricultural Research Service. Office of Pest Management Policy. National Plant Disease Recovery System (2012). Available online at: http://www.ars.usda.gov/Research/docs.htm?docid=14271.
26. U.S. Department of Agriculture. Food Safety and Inspection Service. Food Defense Plan: Security Measures for Food Defense (2009). Available online at: http://www.fsis.usda.gov/PDF/General-Food-Defense-Plan-9-3-09%20_2_.pdf.
27. U.S. Department of Agriculture. Food Safety and Inspection Service. Developing a Food Defense Plan for Meat and Poultry Slaughter and Processing Plants (June 2008). Available on-line at: http://www.fsis.usda.gov/pdf/food_defense_plan.pdf.
28. Extension Disaster Education Network (EDEN). Reducing the Impact of Disasters Through Education (2011). Available online at: http://eden.lsu.edu/Pages/default.aspx.
29. Pravecek, C.L., Davis, J.A., and Greenwood, C.R. Recent Disease Outbreaks and National Exercises. In: Pravecek, Tasha L., Davis, Jim A., and Berg, Thomas (editors). Agroterrorist Attack: DoD Roles and Responsibilities. United States Air Force Counterproliferation Center, Maxwell AFB AL. Chapter 2, pp. 17–31 (2006).
30. National Center for Food Protection and Defense. What is FoodSHIELD? (2012). Available online at: https://www.foodshield.org/index.cfm/discover-tools-links/what-is-foodshield/.
31. Franco, C., and Sell. T.K. Federal Agency Biodefense Funding. FY 2011-FY 2012. Biosecurity and Bioterrorism: Biodefense Strategy, Practice and Science. 9(2): 117–137 (2011). Available online at: http://www.upmc-biosecurity.org/website/resources/publications/2011/pdf/2011-06-07-biodeffunds.pdf.
32. Parker, H.S. Protecting Food and Agriculture: Private Sector Perspectives. Presentation to Georgetown University Medical Center course in agroterrorism (February 4, 2011).
33. U.S. Department of Homeland Security. Agriculture and Food Critical Infrastructure and Key Resources Sector-Specific Plan as input to the National Infrastructure Protection Plan (May 2007). Available online at: http://www.usda.gov/documents/nipp-ssp-ag-food.pdf.
34. U.S. Department of Agriculture. Agricultural Research Service. About ARS (2012). Available online at: http://www.ars.usda.gov/aboutus/aboutus.htm.
35. U.S. Department of Agriculture. Agricultural Research Service. National Animal Disease Center (2012). Available online at: http://www.ars.usda.gov/AboutUs/AboutUs.htm?modecode=36-25-30-00.
36. U.S. Department of Agriculture. Agricultural Research Service. Southeast Poultry Research Laboratory (2012). Available online at: http://www.ars.usda.gov/main/site_main.htm?modecode=66-12-07-00.
37. U.S. Department of Agriculture. Food Safety and Inspection Service. FSIS Laboratories (2012). Available online at: http://www.fsis.usda.gov/science/fsis_laboratories/index.asp.
38. U.S. Department of Agriculture, Animal and Plant Health Inspection Service. National Animal Disease Center (2011). Available online at: http://www.ars.usda.gov/Main/docs.htm?docid=3582.
39. National Center for Foreign Animal and Zoonotic Disease Defense. FAZD Center (2012). Available online at: http://fazd.tamu.edu/.
40. National Center for Food Protection and Defense (undated), Available online at: http://www.ncfpd.umn.edu/.

41. U.S. Department of Agriculture. USDA Opens New Biosafety Level 3 Facility. News release (April 22, 2004). Available online at: http://www.usda.gov/wps/portal/usda/usdahome?contentidonly=true&contentid=2004/04/0157.html.
42. U.S. Department of Homeland Security. National Biodefense Analysis and Countermeasures Center (undated). Available online at: http://www.dhs.gov/national-biodefense-analysis-and-countermeasures-center.
43. U.S. Department of Agriculture. Agricultural Research Service. Foreign Disease-Weed Science Research Unit (2011). Available online at: http://www.ars.usda.gov/main/site_main.htm?modecode=19-20-00-00.
44. Food Safety Summit. Interactive Food Defense Workshop with Government and Industry Representatives (April 17, 2012). Available online at: http://www.foodshield.org/default/assets/File/FD%20Workshop%20Slides_FINAL%20SLIDE%20DECK_5-1-12%5B3%5D.pdf.
45. Farm Foundation. Issue report: Agricultural Research and Productivity for the Future (September 2009). Available online at: http://www.farmfoundation.org/news/articlefiles/105-FINAL%20Sept09_IR%20(2).pdf.
46. Stewart-Brown, B. Agroterrorism Preparedness: Perdue History. Presentation to Georgetown University Medical Center course in agroterrorism (February 11, 2012).

Timeline of Selected Events in Biosecurity Policy

Gigi Kwik Gronvall

1996

The select agent program launched by the Antiterrorism and Effective Death Penalty Act of 1996, which prohibited the transfer of some "select" agents from one laboratory to another without registration with the CDC.[1] Congress passed the law in response to the difficulties in prosecuting Larry Wayne Harris, a member of the white supremacist group Aryan Nations, who in 1995 legally obtained *Yersinia pestis,* the causative agent of plague, from American Type Culture Collection, a biological services company.[2]

2001

After the anthrax letter attacks, a gap in the select agent program was exposed: the CDC maintained a registry of laboratories that had transferred samples of *Bacillus anthracis,* the causative agent of anthrax, but they could not definitively state that those were the only laboratories doing legitimate research on that pathogen.

2003

The Public Health Security and Bioterrorism Preparedness and Response Act of 2002,[3] along with the U.S. Patriot Act of 2001,[4] led to a new select agent program. The major changes from the previous program included:

1. Registration required for all those who have access to select agents
2. Background checks required for all those with access to select agents, performed by the Department of Justice

3. Laboratory inspections by (sometimes joint) teams from USDA (APHIS) and HHS (CDC)
4. Select agent clearance renewal every five years

2003

In October 2003, the National Research Council published Biotechnology Research in an Age of Terrorism, which is known as the Fink Report. The committee described an initial set of seven types of experiments of concern that, while not prohibited, would merit review and discussion before being undertaken or published in detail. Such experiments would do at least one of the following:

1. Demonstrate how to render a vaccine ineffective
2. Confer resistance to therapeutically useful antibiotics or antiviral agents
3. Enhance the virulence of a pathogen or render a nonpathogen virulent
4. Increase a pathogen's transmissibility
5. Alter the host range of a pathogen
6. Enable the evasion of diagnostic and/or detection modalities
7. Enable the weaponization of a biological agent or toxin[5]

In addition to recommending creation of programs to educate scientists about the dual-use dilemma and their responsibility to reduce risks, the committee identified opportunities for review of scientific experiments and results throughout the research lifecycle and called for the creation of the a National Science Advisory Board for Biodefense to propose guidance and leadership for research oversight.[6]

2004

The National Science Advisory Board for Biosecurity was established in 2004 as a federal advisory committee and has been chartered continuously to date at two-year intervals by the Secretary of the Department of Health and Human Services (HHS). The NSABB has offered guidance for handling dual-use research of concern (DURC), a modified form of the Fink Report's experiments of concern, and has provided input on the publication of several papers in addition to the H5N1 papers, including one on the reconstruction of the 1918 influenza virus,[7] for which they recommended publication.

2005

Laboratory accident at Boston University, with one confirmed case and two probable cases of tularemia. The researchers thought they were working with the live vaccine strain (LVS) of *Francisella tularensis,* which is attenuated and should not have caused disease.[8]

2007

Laboratory accidents with *Brucella* and *Coxiella burnetti* at Texas A&M were not immediately reported to the CDC, as was required for all such incidents with select agent pathogens.[9] In response to the violations, there was a congressional hearing in October 2007, which also focused on the expansion of high-containment laboratories.[10] CDC leveled a fine of $1M on Texas A&M, and there was a suspension of select agent work there for many months.[11]

2008

In August, Bruce Ivins committed suicide, and allegations that he was the Amerithrax mailer were revealed to the public.[12] The FBI declared that Bruce Ivins was solely responsible for the anthrax letter attacks on February 20, 2010.[13]

2008

World at Risk, the report of the Congressional Commission on the Prevention of Weapons of Mass Destruction Proliferation and Terrorism (WMD Commission), was released on December 2, 2008. In the report, the WMD Commission recommended that the U.S. should "conduct a comprehensive review of the domestic program to secure dangerous pathogens ... tighten government oversight of high-containment laboratories ... promote a culture of security awareness in the life sciences community."[14]

2009

Working Group on Strengthening the Biosecurity of the United States report released on October 23, 2009.[15] Established by Executive Order 13486,[16] which was signed on January 9, 2009, by George W. Bush, the working group was co-chaired by officials from HHS and DoD. They were tasked to examine the physical/facility security, personnel reliability for handling, storage, or transport of select agents and toxins (federal and nonfederal laboratories). Among their findings and recommendations:

1. The 82 agents on the biological select agents and toxins (BSATs) list do not pose equivalent levels of risk to public or agricultural health, and the list should therefore be either reduced or stratified so that biosecurity measures can be applied according to risk.
2. In response to their finding that institutions reported confusion about inventory management and recordkeeping requirements, they recommended that the relevant agencies provide comprehensive guidance in that area.
3. They recommended improvements to the security risk assessment at the federal level to improve the vetting of U.S. citizens and foreign nationals who enter the select agent program, including continuous monitoring measures once access has been granted.

2009

National Science Advisory Board for Biosecurity (NSABB) Report: Enhancing Personnel Reliability among Individuals with Access to Select Agents released April 30, 2009.[17]

Among the recommendations:

1. Personnel reliability measures, which are designed to counter the threat to biosecurity posed by an insider, should be increased at institutional discretion. However, the NSABB decided that it was not appropriate to have a national program due to possible detrimental effects, including the different levels of risk and resources across public health, university, and federal laboratories.
2. Select agent clearance process should be more stringent.
3. Given the unequal levels of risk of the various select agent pathogens, the select agent list should be reduced or stratified.
4. Regarding dual use research of concern (DURC), which is research that could be misapplied for harm, the NSABB recommended that scientific professional societies strengthen their outreach and education efforts to promote a "culture of research responsibility and vigilance."

2009

The Trans-Federal Task Force on Optimizing Biosafety and Biocontainment Oversight,[18] co-chaired by HHS and USDA, released a report that examined gaps in the oversight framework for laboratories and options for improvement to biosafety/biocontainment oversight. Among its recommendations: there should be training standards and core competencies for personnel at high and maximum containment research laboratories, and there should be a phased-in requirement for credentialing of biosafety professionals at high and maximum containment laboratories.

2009

Defense Science Board task force report Biological Safety and Security Program for Research Involving Biological Select Agents and Toxins publicly released May, 2009.[19] They found that DoD laboratory security for biological elect agents and toxins exceeds the security found in the laboratories of civilian counterparts, especially in addressing the "insider threat." They identified areas for action, including computer security enhancements such as video monitoring of laboratoriy workers, which they determined to be superior to the "two-person rule," transporting pathogens with a "lost in the crowd" approach vs. a two-person escort; and avoiding security measures that are significantly detrimental to the mission, particularly for overseas laboratories.

2009

Government Accountability Organization (GAO) report High Containment Laboratories: Issues related to Oversight released in September 2009.[20] GAO was asked to address the following questions: To what extent, and in what areas, has the number of high-containment labs increased in the United States? Which federal agency is responsible for tracking the expansion of high-containment labs and determining the associated aggregate risks? What lessons can be learned from recent incidents at high-containment labs? The group's findings were that the numbers of high and maximum containment laboratories is increasing and that no one federal agency has the mission to track the expansion of laboratories as well as regulate biosafety practices within them, and it recommended that these areas be corrected. In addition, the group recommended that a mechanism for sharing lessons learned across high containment laboratories be developed. Regarding personnel reliability programs, it recommended that the costs and impacts of such a program be identified prior to establishment.

2009

National Research Council (National Academy of Sciences) report Responsible Research with Biological Select Agents and Toxins Released.[21] Among the recommendations:

1. The list of select agents and toxins should be stratified in risk groups.
2. Accountability measures in the laboratory are best exemplified by controlling access to working materials and stocks, but requirements for counting the numbers of vials or such other required inventory management schemes should not be used because they are unreliable and counterproductive.
3. There should be an appeals process for the security risk assessment screening process for workers to gain access to select agent pathogens.
4. Inspectors of laboratories should have scientific and laboratory knowledge and experience.
5. Regarding personnel reliability screening, the committee was concerned that, because the number of individuals who represent true security risks is likely to be very small, efforts at increasing screening beyond the levels that are employed for people to have access to classified and proprietary information risk increasing the number of innocent individuals who will not pass through security. It advised that there was insufficient information to have routine or random drug testing.
6. Above all, the committee recommended that laboratory leadership be charged with implementing programs and practices aimed at fostering a culture of trust and responsibility alongside ongoing monitoring of personnel issues. Federal agencies overseeing BSAT research should provide educational and training resources to promote effective laboratory leadership.

2009

National Strategy for Countering Biological Threats released November 2009.[22] The strategy includes a series of objectives that are intended to "protect against the misuse of the life sciences to develop or use biological agents to cause harm." These objectives are: Promote global health security (which reduces the impact of infectious disease of all kinds, whether deliberate, natural, or accidental); reinforce norms of safe and responsible conduct (a culture of responsibility to guard against the dual-use problem); obtain timely and accurate insight on current and emerging risks; take reasonable steps to reduce the potential for exploitation; expand the capability to prevent, attribute, and apprehend; communicate effectively with all stakeholders; and transform the international dialogue on biological threats. The emphasis of this strategy was on prevention of biological terrorism and the global nature of the threat.[23]

2010

Executive Order 13546, Optimizing the Security of Biological Select Agents and Toxins in the United States, was released on July 2, 2010.[24] The EO stated that there was an "absence of clearly defined, risk-based security measures" for the select agent program, which "has raised concern about the need for optimized security and for risk management." The EO was also responding to concerns that the cost and complexity of compliance with the select agent program would discourage researchers from remaining in and entering the field. As "research and laboratory work on BSAT is essential to both public health and national security, it is in the interest of the United States to address these issues." Among the components of the EO:

1. It called upon HHS and USDA to develop a risk-based tiering of the biological select agents and toxins (BSAT) and designate a subset of the select agent list (tier 1) that presents the "greatest risk of deliberate misuse with most significant potential for mass casualties or devastating effects to the economy, critical infrastructure, or public confidence." It also called upon HHS and USDA to explore options for graded protection of tier 1 agents and to consider reducing the overall number of agents and toxins on the select agent list.
2. It created the federal experts security advisory panel (FESAP) to make technical and substantive recommendations on BSAT security concerning the SAP. The FESAP consisted of members from a variety of government agencies, including HHS, USDA, Department of Defense (DoD), Department of Justice (DOJ), Environmental Protection Agency (EPA), Department of Energy (DOE), and many others.

2011

On June 13, 2011, the federal experts security advisory panel released Recommendations Concerning the Select Agent Program,[25] as called for in EO 13546.[24] It created a tier 1 of pathogens and recommended that 25 pathogens be removed from the select agent list. Regarding personnel reliability, the panel "will further explore the utility of behavioral assessments to identify indicators of potential for violent behaviors, criminal behaviors, or other behaviors that pose a national security risk" and recommended enhancements to employee monitoring. In contrast to the many reports that had recommended a stratified approach to risks regarding select agent pathogens and less security for many pathogens on the list, FESAP took current select agent security measures as a baseline, and increased controls on tier 1 pathogens:

1. The FESAP report: consists of expert members from 15 departments, agencies, and offices and is co-chaired by Dr. Gerald Parker, HHS, and Dr. Gregory Parham, USDA. (On December 20, 2010, Dr. George Korch replaced Dr. Parker as the HHS co-chair.)
2. Tier 1 includes Bacillus anthracis, Burkholderia mallei, Burkholderia pseudomallei, ebola virus, foot and mouth disease virus, Francisella tularensis, Marburg virus, variola major virus, variola minor virus, and Yersinia pestis
3. The panel also recommended the removal of a number of animal and human pathogens from the select agents list and recommended against the inclusion of SARS,
4. Behavioral assessment: "As part of its continuing work, FESAP will further explore the utility of behavioral assessments to identify indicators of potential for violent behaviors, criminal behaviors, or other behaviors that pose a national security risk" and credit checks: "If possible under current statutory authority, the SAP should require credit checks to be performed by the entity to assess financial responsibility and should provide guidance on procedures for utilizing the information as part of a suitability assessment." SAP should provide guidance for self- and peer-reporting for safety and security concerns.

References

1. The Antiterrorism and Effective Death Penalty Act 110, Stat. 1214 (1996).
2. Preston, R. Annals Of War—The Bioweaponeers. The New Yorker (pp. 52–65) (1998).
3. Public Health Security and Bioterrorism Preparedness and Response Act (2002).
4. Uniting and Strengthening America by Providing Appropriate Tools Required to Intercept and Obstruct Terrorism (U.S. Patriot Act) Act. Sec. 817 (2001). Expansion of the Biological Weapons Statute (2001).
5. National Research Council (U.S.). Committee on Research Standards and Practices to Prevent the Destructive Application of Biotechnology. Biotechnology research in an age of terrorism. Washington, DC. National Academies Press (pp.5–6) (2004).
6. National Research Council (U.S.). Committee on Research Standards and Practices to Prevent the Destructive Application of Biotechnology. Biotechnology research in an age of terrorism. Washington, DC. National Academies Press. p. 3 (2004).
7. Kennedy, D. Better never than late. Science. 310(5746):195 (October 14 2005).
8. Barry, M.A. Report of Pneumonic Tularemia in Three Boston University Researchers (March 28, 2005).
9. CIDRAP News. CDC details problems at Texas A&M biodefense lab (September 5, 2007).
10. Germs, Viruses, and Secrets: The Silent Proliferation of Bio-laboratories in the United States. Committee on Energy and Commerce. Available online at: http://www.gpo.gov/fdsys/search/pagedetails.action?st=Laboratory+Attack&granuleId=CHRG-110hhrg44948&packageId=CHRG-110hhrg44948&bread=true2007.

11. Kaiser, J. Texas A&M to Pay $1 Million for Biosecurity Breaches. Science Now (February 20, 2008). Available online at: http://news.sciencemag.org/sciencenow/2008/02/20-01.html (accessed July 17, 2012).
12. U.S. officials declare researcher is anthrax killer. CNN. (August 6, 2008). Available online at: http://articles.cnn.com/2008-08-06/justice/anthrax.case_1_bruce-ivins-anthrax-killer-anthrax-attacks?_s=PM:CRIME (accessed June 11, 2012).
13. Serrano, R.A. U.S. closes case on anthrax letters. L.A. Times (February 20, 2010). Available online at http://articles.latimes.com/2010/feb/20/nation/la-na-fbi-anthrax20-2010feb20 (accessed July 17, 2012).
14. Graham, B., Talent, J.M., and Alison, G.T. Commission on the Prevention of Weapons of Mass Destruction Proliferation and Terrorism (U.S.). World at risk : the report of the Commission on the Prevention of WMD Proliferation and Terrorism. Washington, D.C. Supt. of Docs. U.S. G.P.O. (2008).
15. Report of the Working Group on Strengthening the Biosecurity of the United States (October 23, 2009). Available online at: http://www.phe.gov/Preparedness/legal/boards/biosecurity/Documents/biosecreportfinal102309.pdf (accessed July 18, 2012).
16. President George W. Bush. Executive Order 13486—Strengthening Laboratory Biosecurity in the United States (January 9, 2009). Available online at: http://www.gpo.gov/fdsys/pkg/WCPD-2009-01-19/pdf/WCPD-2009-01-19-Pg32.pdf) accessed May 15, 2012).
17. National Science Advisory Board for Biosecurity. Enhancing personnel reliability among individuals with access to Select Agents (2009). Available online at: http://oba.od.nih.gov/biosecurity/meetings/200905T/NSABB%20Final%20Report%20on%20PR%20 5-29-09.pdf (accessed July 18, 2012).
18. Report of the Trans-Federal Task Force on Optimizing Biosafety and Biocontainment Oversight (July 2009). Available online at: http://www.ars.usda.gov/is/br/bbotaskforce/biosafety-FINAL-REPORT-092009.pdf (accessed July 18, 2012).
19. Report of the Defense Science Board Task Force on Department of Defense biological safety and security program. Washington, D.C. Office of the Under Secretary of Defense for Acquisition, Technology, and Logistics (2009).
20. Government Accountability Organization. GAO-09-574. High-containment laboratories: national strategy for oversight is needed (September 2009). Available online at: http://www.gao.gov/new.items/d09574.pdf (accessed July 18, 2012).
21. National Research Council (U.S.). Committee on Laboratory Security and Personnel Reliability Assurance Systems for Laboratories Conducting Research on Biological Select Agents and Toxins. National Academies Press (U.S.). Responsible research with biological select sgents and toxins. Washington, D.C. National Academies Press (2009). Available online at: http://site.ebrary.com/lib/pitt/Doc?id=10367629.
22. National Security Council. National Strategy for Countering Biological Threats (November 2009).
23. Koblentz, G.D. From biodefence to biosecurity: the Obama administration's strategy for countering biological threats. International affairs.88(1):131–148 (2012).
24. President Barack Obama. Executive Order 13546—Optimizing the Security of Biological Select Agents and Toxins in the United States (July 2, 2010). Available online at: http://www.whitehouse.gov/the-press-office/executive-order-optimizing-security-biological-select-agents-and-toxins-united-stat (accessed May 15, 2012).
25. Federal Experts Security Advisory Panel. Recommendations Concerning the Select Agents Program (June 13, 2011). Available online at: http://www.phe.gov/Preparedness/legal/boards/fesap/Documents/fesap-recommendations-101102.pdf (accessed May 15, 2012).

Index

A

Academic institutions, 51, 83
Access control, 19, 22, 57, 58, 60, 64, 68, 95
Accountability, 18, 72, 75,
Administrative controls, 18, 72, 95, 261, 267
Adversaries, 58, 91, 93
Agricultural infrastructure, 131
Agroterrorism, 34, 133, 135, 138
Animal Health Protection Act of 2002, 76, 272
Animal Plant Health Inspection Service (APHIS), 19, 84
Anthrax, 8, 19, 33, 36, 59, 115, 134, 157, 160, 161, 188, 261
Asian soybean rust, 140, 275
Attack cycle, 197, 198, 203
Aum Shinrikyo, 30, 33
Autonomous detection technology equipment, 162

B

Bioaccidents, 37
Biocontainment, 5, 7, 237, 263
Biocrime, 34, 36
Biodefense, 5, 8, 273, 279
Biohazard, 5–6, 8, 250–251
Biological aerosol sentry and information system (BASIS), 161
Biological select agents and toxins (BSAT), 17, 53, 92, 118, 188, 202, 263, 292, 293, 295
Biological warfare, 28, 31
Biorisk assessment, *see risk assessment*
Biorisk management, 5, 22, 220, 239, 262
Biorisk, 5, 21, 114
Biosafety, 6, 17, 20, 21, 39, 56, 91, 114, 168, 169, 172, 225, 226, 227, 230, 238, 260
  culture, 23
  in microbiological and biomedical laboratories (BMBL), 18, 72, 103, 114, 116, 239, 262
  officer, 83, 267
Biosecurity
  policy, 187, 261, 289
  policy statement, 75, 76
  program plan, 72, 77
Biosurety, 8, 102, 190, 225, 226, 237, 261
Biosurveillance (disease surveillance), 162, 180, 189, 225, 237, 264, 276, 280
Bioterrorism, 11, 30, 89, 133, 197
  attack cycle, 197, 198
BioWatch program, 161

Bioweapons, 9, 11, 135, 170, 205
Boston University, 291
Bruce Ivins, 92, 117, 189, 291
BSL-3, 37, 113, 226, 260, 281
BSL-4, 37, 113, 190, 226

C

Centers for Disease Control & Prevention (CDC), 4, 6, 17, 27, 158, 178, 188, 225, 282
Chain of infection, 171
Citrus canker, 141
Codes of conduct, 95, 212, 214
Community, 23, 92, 115, 168, 197, 215, 219
  academic, 216
  agricultural, 174, 276
  biodefense, 262
  biosecurity, 14, 264
  biomedical, 90, 92
  international, 180
  local, 8
  police, 102, 115
  public health, 90, 98, 105, 122, 174
  research, 38, 141
  scientific, 13, 119–120, 122, 175, 191, 210, 216, 219–221, 291
  security, 117, 119, 175, 215, 217, 220
Countermeasures, 89, 102, 155, 161, 197, 202, 203, 207

D

Department of Health and Human Services (HHS), 148, 188, 274, 290
Detection, 58, 61, 66, 100, 155, 161–162
Dual-use, 113, 118, 170, 187, 191, 209, 212, 264, 290

E

Emergency response, 64–65, 67, 158
  efforts, 8
  network, 275
  personnel, 160
  plan(s), 18–19, 156, 158, 275
  procedures, 73
  team(s), 78, 81
  training, 72
Engagement and education, 209
Epidemic(s), 28, 33, 167, 168, 171–173, 176–180, 263
  capability, 29
Events
  public health, 158–159
Executive Order 13486, 117, 291
Extension disaster education network (EDEN), 278

F

Federal Experts Scientific Advisory Panel (FESAP), 190
First response, 59, 263
Food and Agriculture Organization (FAO), 167, 230, 263
Food emergency response network, 275
Food Safety Modernization Act of 2010, 272, 277
Food security, 130, 282
FoodSHIELD, 278
Foot and mouth disease, 133, 138, 177, 180, 278, 280, 296

G

Global Early Warning, 189
Global health, 259, 261, 294
Government Accountability Organization (GAO), 293
Government facilities, 83, 280

H

H5N1 Papers, 113, 119, 120, 192, 290

I

Immune response, 172
Incident response, 37, 73, 81, 86, 277
  plan, 73, 76, 81
Influenza
  avian, 120–121, 178–179
  H5N1, 120–121, 191
  in 1918, 189, 290

Information security, 81, 180, 226, 235, 237
Insider threat, 58, 73, 89
International health programs, 178

K

Key control, 68

L

Laboratory
- acquired infections (LAI),12, 21, 37, 173
- biosafety, 226
- design considerations (A/E), 63

Law enforcement outreach, 213, 214
Layered security, 57, 65
Logic model, 228, 231

M

Mass transit, 155
Melamine contamination, 142
Mitigation, 52, 77, 100, 277

N

National Academies of Science (NAS), 103, 190, 213
National bio- and agrodefense facility (NBAF), 281
National Biosurveillance Integration Center (NBIC), 276
National Infrastructure Protection Plan, 274, 277, 279, 283
National Interagency Confederation for Biological Research (NICBR), 282
National Science Advisory Board for Biosecurity (NSABB), 103, 118, 191, 290, 292

O

Occupational health programs, 23, 227
One World, One Health, 262
Outbreaks, 9, 34, 135, 139, 168, 172, 262, 278, 282
Outbreak response, 174–175, 179
Outreach, 119, 213, 214, 216, 278
Outsider threat, 59, 73, 207, 261

P

Patriot Act, 19, 76, 104, 115, 188, 289
Personnel reliability, 8, 22, 118, 190, 291, 292, 295
Physical biosecurity, 13, 51, 54
Plum Island Animal Disease Center, 94, 272, 280, 281
Policy, 76, 118, 187, 191, 259, 261, 271, 289
Preemptive measures, 97–98
Primary containment, 7, 237, 238
Principal component analysis, 229, 230, 244, 246, 247, 248
Private organizations, 84
Public Health Security and Bioterrorism Preparedness and Response Act, 19, 104, 188, 289
Public health, 99, 121–122, 167–175, 179, 215, 227–228, 230, 259, 263–265, 276, 282
- community, 90, 98, 105, 122, 174
- definition of biosafety in, 169, 259
- epidemics, 178
- emergency, 227
- events, 158–159
- institutions, 90
- negatively impact, 5
- researchers, 170
- risk, 115, 120, 172–173, 292
- security, 170, 175, 295
- surveillance, 225–226, 238, 263
- threat, 92, 191
- workers, 63, 169, 265

Q

Quality assurance, 69

R

Rajaneeshees, 33–34
Recovery, 58, 63, 141, 144–145, 148–149, 156, 158, 163, 198, 272–272, 277

Regulation(s), 54, 60, 75–77, 79, 81–85, 103–105, 107–108, 118, 170, 187, 189, 191–193, 227, 244, 249, 260
Research laboratories, 14, 17, 27, 37, 113–123, 261, 263, 293
Response, 58–65, 67, 69, 85, 101, 140, 143–145, 169, 171–172, 174–176, 178, 227, 275, 277
  capabilities, 56–57, 63, 117
  drill, 157
  effort, 163
  emergency, 64, 65, 67, 72, 78, 81, 158, 275
  emergency personnel, 160
  emergency plan, 8, 18–19
  first, 59, 263
  force, 61
  immune, 172
  incident plan, 37, 81
  outbreak, 174–175, 179, 181, 263
  plan, 73, 76, 81–82, 86, 100–101, 155–156, 158–159, 161, 265, 272–273, 275, 277
  procedures, 37, 69, 73
  program, 147
  protocol, 81
  team, 82
  time, 62, 85
  training, 72, 86
Risk assessment, 5, 12–14, 19, 21–24, 54–56, 62, 72–24, 76, 80–82, 86, 91, 104, 108, 116, 118, 143, 158, 219, 240–241, 260–261, 266, 276, 283, 292, 294
Risk management, 5–6, 12, 14, 21-24, 82, 114, 117, 168, 220, 239, 262, 274, 276, 284, 295

S

Secondary containment, 7, 238
Security plans, 12–13, 67–68, 102, 116, 237, 261, 283
Select agent program, 19, 103, 105, 115-118, 188-189, 191, 289, 292, 295
Standard operating procedures (SOP), 13, 54, 75, 78, 178, 181
Strategic National Stockpile (SNS), 160
Suitability programs, 97, 99, 102–106, 266
Sustainability, 219, 230, 282
Sustainable design and operations, 67
Synthetic biology, 170, 192–193, 200, 205, 210, 217

T

Targeted violence process, 74, 92, 95–98, 100, 266
Texas A&M University, 116, 281
Threat, 6, 9, 13–14, 18, 21, 24, 27, 30, 34-35, 52, 55–62, 68, 72–75, 77, 81–84, 113, 117–120, 123, 133–135, 141, 145–146, 155–156, 158, 161–162, 170-171, 189–191, 203, 207, 219, 237, 272–274, 179, 287, 292–294
  assessment, 14, 22, 55–56, 58, 60, 72–74, 89, 91–93, 95–98, 100, 102–108, 115, 260–261, 266
  external, 74, 95–96, 106, 115, 118
  indirect, 91, 95–96
  internal, 95–96
  public health, 92, 191
Training, 35, 37–38, 59, 67–69, 75–76, 79–80, 83, 85–86, 90, 100, 104, 114–115, 148, 177, 181, 190, 193, 198–199 209-220, 229, 238–239, 244, 249, 263, 275, 277, 280, 284
  biosecurity, 72, 76, 267
  emergency response, 72
  events, 75
  exercises, 69, 141
  first-aid, 23
  needs, 56
  manual, 241
  model, 236
  outcomes, 75
  preparedness, 23
  procedure, 69
  programs, 148, 273

resources, 294
response, 86
standards, 293
Trans-Federal Task Force on Optimizing Biosafety and Biocontainment Oversight, 37, 292

U

U.S. Department of Agriculture, 19, 23, 38, 132, 140, 142, 146–148, 150, 169, 188, 271–282, 290–291, 295–296
United States Army, 30, 36, 189
United States Army Medical Research Institute of Infectious Diseases, 36, 92, 117, 189
USAMRIID, *see United States Army Medical Research for Infectious Diseases*
USDA, *see U.S. Department of Agriculture*

V

Valuable biological material (VBM), 3, 21, 53, 56, 76, 87, 89–90, 168, 170, 200, 205–207, 230, 237
Vulnerability assessment, 14, 55–57, 67, 273, 276, 283

W

Weapons of Mass Destruction Proliferation and Terrorism (WMD Commission), 117, 291
World Animal Health Information Database (WAHID), 189
World Health Organization Laboratory Biosafety Manual, 7

Printed and bound by CPI Group (UK) Ltd, Croydon, CR0 4YY
16/09/2026
14963417-0002